The Challenge of Change

The Challenge of Change

Report of a conference on Technological Change and Human Development at Jerusalem, 1969

TREVOR EVANS

1966

PERGAMON PRESS

Oxford · London · Edinburgh · New York
Toronto · Sydney · Paris · Braunschweig

Pergamon Press Ltd., Headington Hill Hall, Oxford
4 & 5 Fitzroy Square, London W.1
Pergamon Press (Scotland) Ltd., 2 & 3 Teviot Place, Edinburgh 1
Pergamon Press Inc., Maxwell House, Fairview Park, Elmsford, New York 10523
Pergamon of Canada Ltd., 207 Queen's Quay West, Toronto 1
Pergamon Press (Aust.) Pty. Ltd., 19a Boundary Street,
Rushcutters Bay, N.S.W. 2011, Australia
Pergamon Press S.A.R.L., 24 rue des Écoles, Paris 5e
Vieweg & Sohn GmbH, Burgplatz 1, Braunschweig

First edition 1970
Library of Congress Catalog Card No. 71-104788

Printed in Great Britain by The European Printing Company, Bletchley

08 015824 2 (flexicover
08 015825 0 (hard cover

Contents

Foreword

by the

MINISTER OF TECHNOLOGY
*Millbank Tower London SW*1

TOMORROW'S world has already arrived. The moon landings are the most dramatic demonstration of this fact, but the impact of technology has, in a thousand less spectacular ways, revolutionized the pattern of living within a single generation.

Since the beginnings of civilization, the pace of technological change has allowed plenty of time for man to adjust himself to these advances. Today the rate of change has so accelerated that we are all like men who fly to Australia and discover that our body has not had time to catch up with the new time-scale.

There is just no longer the time for leisurely adjustment. If we are going to get the best out of technology we must make a continuing conscious effort to meet change by studying the effect it has on the world, on governments, on institutions, and, most important of all, on the individual.

The Jerusalem Conference was an important contribution to this sort of thinking. Trevor Evans's report is intensely readable and will help us all to focus attention on to the problems and the opportunities which technology presents to the world.

Anthony Wedgwood Benn

Foundation on Automation and Employment Limited,

13 Southampton Place,

London, W.C.1

Council of Management

Chairman

The Rt. Hon. Lord Carron, KSG, MA, DSc

Vice-Chairman

The Rt. Hon. Lord Robens of Woldingham, PC, DCL, LLD

Secretary-Treasurer

The Rt. Hon. Lord Hirshfield, FCA

Sir Leon Bagrit

Leslie Cannon, CBE, MA

Michael Clark

The Rt. Hon. Lord Cooper of Stockton Heath, JP

The Rt. Hon. Lord Douglass of Cleveland

Sebastian de Ferranti, DSc

Sir Sidney Ford, MBE

Maurice G. Kendall, ScD

Sir Joseph Lockwood

Edwin R. Nixon

Sir Reginald Pearson, OBE

Cyril Plant, OBE

Professor B. C. Roberts, MA

Edward Senior, CMG

Douglas Taylor, CMG

Sir Charles Wheeler, KBE

Director

B. John A. Hargreaves, JP

Chapter 1

New Courage in a New World

A SLIGHTLY cynical, but highly successful, friend was intrigued to see me at London Airport one Sunday morning in April 1969 in a queue which included some eminent businessmen, trade union leaders, and academics. And we were bound for Tel Aviv. Still, he asked me where we were going, and I said "Jerusalem".

He asked me why? And in a consciously pompous way, I intoned, "The International Conference on Technological Change and Human Development", at which he whistled sardonically, shrugged his shoulders, and repeated the war-time inquiry, "Is your journey really necessary?" This report is the answer to him and to sceptics everywhere.

True, the formal title of the conference might be forbidding. But there can be few assemblies anywhere more distinctive than one dedicated to discuss the adaptability of man to the marvels of his own creation. If man fails to respond those marvels will be regarded as inhuman monsters.

There may be some dubiety on the timing of such a conference. There are conflicting philosophies in the world; there are startling contrasts in material wealth, not only between nations, but quite as markedly in classes within a single nation; there are now, perhaps more strikingly than ever before, animosities between the generations; and the inducements to the technocratic sectors of the advanced nations to enhance their superiority could make them callously indifferent to the resignation and the hopelessness of unskilled workers or of underdeveloped countries.

Not that there were any such misgivings on the part of the sponsors of the conference, the Foundation on Automation and

Employment Ltd. (Britain), the American Foundation on Automation and Employment, Inc., and the Israel Steering Committee. On the contrary, they felt, as did virtually all the 170 or so participants at the conference, that the progress of technology is assured and inevitable and accelerating, but the future of the social institutions of man cannot be predicted with such confidence without the most assidous preparation for his intellectual ability to retain his control of technology, and the protection of those whose outmoded crafts or traditional ways of life are bound to be displaced by modernity. There is no time like the present. Delay would merely exacerbate the problems.

Even the venue of the conference created some surprise. Why, of all places, Jerusalem?

This is one of the historic cities of the world, a cradle of faith, a stimulus to emotion, bizarre in the contrast of the visible evidence of past glories and the energetic pursuit of wisdom to create a prosperous future. There is a determination to create a state of rare egalitarianism, one in which, in the simple moving words of Yeruham Meshel, one of the leaders of the Israeli trade unions, "it would be difficult to be too poor, and difficult to be too rich". The handicaps to such an ambition are tremendous in a country which is isolated on its landward sides by a sea of hatred, forced to spend between a quarter and a third of its wealth on defence and much of its time on vigilant security.

But these were not the considerations which led to the selection of Jerusalem as the conference centre, even though their existence made a profound impression on the participants. There was a shrewder and more appropriate reason for including Israel within the triumvirate of nations sponsoring this conference. As Britain in the Western world is the oldest of the industrial nations and the United States the most prosperous, so is Israel the newest and, possibly, the most exciting.

Four times daily a fleet of coaches, emblazoned with the name of the conference across their bonnets, would collect visiting participants from three of the city's leading but scattered hotels to deliver them to and return them from the two sessions in the well-equipped Wise Auditorium of the Hebrew University. Not infrequently, workers on the new roads and on the building sites, the shoppers in their queues, and even strollers passing by would stop and wave with enthusiasm. There was a sense of appreciation, of participation. This is a land where technology is more respected than feared. It is also a land where there is an acute realization, from government ministers to taxi-drivers and waiters, that Israel is a back-marker in its bid to be included among the advanced nations of the world, and it is not enough to avoid the past mistakes of more experienced industrial nations. The greatest aid will come from embracing and adapting the latest techniques.

Meticulous care had been taken in the preparation of the conference programme to ensure a strict division of responsibility to each of the three nations. Thus, of the twelve prepared papers which were the bases of the discussions, four each came from British, American, and Israeli contributors. Although the discussions were free and open and well supported by a wide variety of participants, there was, on some topics, a natural tendency for groups to respond to submissions from their own compatriots.

There was, however, ample evidence of an absence of any chauvinistic or nationalistic boastfulness. There were three main reasons for this. The constitution of each group included representatives of government, employers, trade unions, and academics. There was thus a "horizontal" element in which, for example, the employers in any country would be particularly interested in the experiences and intentions of their counterparts elsewhere; trade unionists have an absorbing interest in the variations of union practices everywhere; and academics have a natural instinct for hearing the views of others even when they are accustomed to expressing their own.

Secondly, some of the most provocative contributions dealt with international problems far beyond the confines of any one state. Notable examples came from Professor Nicholas Kaldor, Cambridge, who discussed the role of modern technology in raising the economic standards of underdeveloped countries; Mr. Peter Parker, who threw new light on the emergence of international companies; and Professor J. Rodney Crossley, University of Leeds, whose thesis on incomes policy and the sharing of capital gains broke new ground and will doubtlessly be developed and amplified in many future conferences.

Thirdly, the problems of the future are far more important than any satisfaction from the achievements of the present. There is more to worry than to boast about.

In fact, one of the outstanding features of most of the assessments and discussions was the recognition of present shortcomings in all countries. Each country obviously has its own exclusive problems as well as the more general and universal anxiety to instil a more acute sense of awareness on the speed and potential of technological development. Accompanying this was the realization of the magnitude of the measures needed to equip personnel to face-up to the demands of a modernized industrial and commercial world, either in preparing youth for its entry into this world or in adjusting older people to fit into new callings and altered circumstances.

Certainly, there was no complacency, nor did any speaker claim speciously that present conditions were well-developed enough to make many of these transitions easy. Nor, on the other hand, was there gloom and a spineless acceptance of ultimate defeat.

There was, in fact, a healthy sense of realism, and there were vivid reminders of the pressures of present problems. These actually impinged on conference arrangements by causing the departure or absence of scheduled participants. Mr. Whitney Young, for example, the prominent Negro civil rights leader, had to return to the United States after the first session of the conference. Sir Denis Barnes, the Permanent Secretary of Britain's Department of Employment and Productivity, who had prepared one of the most perceptive analyses of the present state of organizations in Britain, was so preoccupied with the controversy between his government and the trade unions that he had to cancel his visit, though the loss to the conference proceedings was considerably minimized by his admirable substitute, Mr. Frank

Pickford. And Professor John Dunlop, who had been assigned the arduous responsibility of formally summing-up the proceedings at the conclusion of the conference, was absent because of a strike at Harvard University.

Some participants were more ready than others to disclose the domestic anxieties of their homelands. And in this respect, the American spokesmen were outstandingly candid.

Mr. Bayard Rustin, a distinguished spokesman for the Negro cause, for example, was obviously acutely aware of the embarrassment of recent signs of anti-Semitism among some groups of Negroes, particularly in New York, and very conscious that the majority of his audience were Israelis.

He gave the assurance: "There has been no group of Americans who have been more dedicated to justice and freedom in the United States than the Jews, who have over the years made contributions to the struggle of the black people in our country for freedom. It would be impossible for me to outline them all, but I would remind you that the first foundation set up in the United States for the purpose of educating Negroes was established by a Jew. And in our struggles over the years there have been Jews who have been jailed, who have been murdered standing with us in Mississippi, and although we have some minor anti-Semitism among black people in the United States, the great majority of our black people recognize the vast contribution of the Jews. Those of us who are in a position of some limited influence will not rest until we have rooted out any last evidence of anti-Semitism, not only because we know that the Jews have made a great contribution, but because we know that so long as an atmosphere is created in which any of God's children are having fingers pointed at them for being inferior or different, given the social and economic structure in the United States, that it will be those of us who are black who will be most brutalized."

Understandably, those American participants preoccupied with the plight of those described as the underprivileged and dispossessed were mainly concerned with Negroes, though one or two pointed out that, numerically, there are more "poor

Whites" in the United States than Negroes. Most United States speakers, by the way, referred to the Negroes quite unequivocally as "Blacks", a pejorative term to British ears, but it was explained privately that it was not generally resented in the United States where some accepted it as the abbreviation of a campaign cry that "Black is beautiful".

Certainly the Negroes had some passionate champions at this conference, but their advocacy did not preclude an occasional wry remark. There was, for example, the vivid recollection of Dr. Kenneth Clark, who wants all the underprivileged to be treated as first-class citizens. There was no doubt of his zeal.

But, he recalled, "In Watts, California, I spent a day with a young man about 22 years old. He was very successful—I knew he was successful because he owned a long black Cadillac and filled his tank frequently with expensive gas. Butch was on three anti-poverty payrolls in addition to his basic employment. And I must tell you that I was not shocked or surprised. I was not susceptible to such middle-class responses. It seemed to me that his behaviour was the personification of the reality of ghetto psychology. It did, of course, merely reflect the more smooth, more privileged, suburban reality of exploitation of situations. Ghetto youth often reveal penetrating and probing insights into the moral erosion that is norm in the larger society."

Dr. Clark was not blaming Negro exploitation. He was merely illustrating his conviction that the anti-poverty programme so far introduced in the United States, though well-intentioned, has failed to get to the roots of ghetto resentment and has legitimized "hustle". "The immorality of the ghetto", he argued, "must be seen as merely part of a larger immorality of a society which tolerates or pretends it can function effectively while millions of human beings are subjected to dehumanizing conditions."

And Mr. Harry Van Arsdale, one of the liveliest American trade union leaders, could not resist the comment, "Well, of course, a definition of poverty is a very important thing for people of different nations to be discussing. In our ghetto areas one of our biggest problems is that we do not have enough parking space for cars! I don't know if that exists in all parts of the world."

Of course, Mr. Van Arsdale was right to draw attention to this definition of poverty, because frequently when American spokesmen were deploring and quoting what they regarded as abysmally low rates of pay, it was almost obvious to see citizens of other countries mentally exchanging these rates into their own currencies and marvelling at the concern of the Americans.

Far more geared to their own experience, especially in Britain, was an anecdote related by Dr. Isador Lubin, economic consultant to America's Twentieth Century Fund.

When discussing the importance of training people for modern jobs, Dr. Lubin pointed out that such people would also have to acquire certain work habits. He explained: "We live by the clock in modern society, and a person is not only going to have to learn the skill but also the way of conforming to accepted practices. People who have never held a job have to undergo a change in their habits."

Dr Lubin recalled: "I think of a friend of mine who ran a factory in Puerto Rico, hiring Puerto Ricans who had never worked in a factory. He found that the absenteeism and the late coming were ruining the plant, which was running on piecework. He made a study of the days when tardiness was greatest and absenteeism at its height, and found it was Monday. He announced on Friday that half-an-hour before the plant opened on Monday, there would be a bingo game with a 50 dollar prize. Everybody turned up. He actually saved the prize from increased production. That happened many times. Devices of that sort can be important."

These examples are recorded, however, merely as examples of relaxed forms of American candour. They were refreshing deviations from the normally deeply earnest and serious tenor of American contributions.

In the main, too, the Israeli participants were preoccupied with the complexities of their transformation into a well-integrated technological state. They begin with the immense advantage of recognizing the supreme importance of education in all its forms. The anxieties of their present defensive state are compensated by an overwhelming realization of the value of co-operation from practically all sectors of the community. They accept restraints which were tolerable in Britain only in a period of war. There is a great deal of speculation among the Israelis on whether this cohesion can be maintained if and when peace comes to the Near East.

Undoubtedly, the Israelis succeeded in giving the impression that they are aware of the magnitude of their present and future tasks, and they are proud of the progress they have made so far. Even such shortcomings as the integration of illiterate immigrants and the shortage of experienced managerial types, especially in the middle ranges of supervision, fail to daunt them.

They, too, contributed their fair quota of stimulating observations. For example, at the conclusion of the discursive discussion stimulated by the wide range of topics covered by Sir Denis Barnes on the roles of government, employers' organizations, and trade unions, the chairman of the session, Professor Jay Tabb, one of the liveliest members of the Israel Institute of Technology, startled the conference when he observed: "I am reminded of the man who, passing a church, saw a sign which read, 'Jesus is the answer', and he remarked, 'Yes, I know, but what is the question?'"

And Mr. Yeruham Meshel, one of the top executives of Histadrut, the Federation of Trade Unions, wondered whether the trade union leaders of Italy and Germany might not have been wiser if they had given stronger support to the protesting students of their country. But he was later to speculate on what might happen if students in Israel were given a chance every 4 years of electing their own professors. He was goaded into this notion by a clinically detached assessment of the shortcomings of trade unions by the diminutive Dr. Rivka Bar-Yosef, chairman of the Department of Sociology at the Hebrew University of Jerusalem.

Perhaps the most disconcerting intervention by an Israeli commentator arose from an idea put up by Professor Nicholas Kaldor, who had suggested in his paper on how underdeveloped countries could be helped towards advancement that in some circumstances they would be wise to purchase second-hand equipment from the more advanced countries. This had provoked a fair amount of dissent.

Staunchly, Mr. A. Hadar assured the conference that Israel had never encouraged the purchase of second-hand equipment. He explained: "Our policy was exactly the opposite. If the volume had not permitted us to bring in new equipment which would enable us to embark on mass production, we confined ourselves to new equipment with a limited output. But as we began to market more, we have passed to mass production processes and now we have the possibility of having more modern and up-to-date equipment."

This was not allowed to go unchallenged by Mr. Larry Leeds, New York, the president of Manhattan Industries, Inc., who observed: "When we talk about second-hand equipment in Israel, I want to remind you that the whole transportation system of this country in its early years was built on second-hand equipment. Every old taxi-cab from New York City was brought over here and performed wonderfully; in fact if you go into the streets you may still see some of them. It was the same with a lot of the aircraft—the old DC3's provided the background to much of the air transportation of this country during its developing stage. I think we have to be very selective when we discuss second-hand equipment—it makes a lot of sense in certain areas."

It can be added that there was no resentment among the Israeli participants at this correction. On the contrary, the whole atmosphere of the conference was one of amiability in the acceptance even of conflicting views, and there was a deep interest in and examination of criticisms.

But one of the shrewdest observations of all came towards the end of the conference, too late to evoke subsequent comments, from Mrs. Mary Conway Kohler, of New York. She commented:

"The British are so reserved, they don't expose their soiled linen quite like we do, or as the Israelis have." Only a little group of British participants chuckled when they overheard one of their compatriots murmur "We've got a lot to be reserved about".

Nevertheless, Mrs. Kohler had raised an intriguing point, for it was a fact that there had been a marked absence of self-criticism among the British participants, even though they had contributed some of the most significant papers and had dominated many of the discussions. In a similar conference in Rome, 2 years earlier, some of the British participants had been so outspoken in their condemnation of some facets of British institutions that they had imposed a curb on themselves towards the closing sessions of that conference to avoid giving the impression that everything was wrong in their homeland.

Mrs. Kohler had, however, provoked them into a series of sporadic and informal self-examinations. They accepted that, on the face of it, Mrs. Kohler's remark was justified, although they pointed out that some of Sir Denis Barnes's conclusions on the weaknesses of employers' organisations and of the TUC were unusually outspoken for a civil servant.

The most frequently expressed explanation for the critical reticences of the British contingent was that the contentious problems of greatest interest to them could be of only a limited appeal to others. Clearly this was a reference to the controversy between the Prime Minister and Mrs. Castle and the TUC on the proposed introduction of trade union legislation. It so happened, too, that Mr. Roy Jenkins, the Chancellor of the Exchequer, was presenting his budget on the second day of the conference—another issue of concern to the British, but to few others.

One of the most interesting innovations at this conference was a "panel session" at its conclusions in which spontaneous questions on any issue raised during the week could be put to a panel of eight experts. Mr. Leslie Blakeman, now a member of Britain's new Commission on Industrial Relations and formerly a Ford (England) executive, presided over the panel, which included three from Britain, three from the United States, and two from Israel. [See Appendix A. p. 176]

Obviously, the main issues raised were the speed of technological development, education, the role of government in collective bargaining, the role of international companies, and the effect of change on poverty. Each of these themes is dealt with in subsequent chapters. Most of the answers from the two employers, the two trade unionists, and the four academics on the panel provided admirable amplifications of the debates which had taken place earlier in the conference.

The most intricate task of the whole week devolved on Mr. Theodore Kheel, president of the American Foundation on Automation and Employment. He delivered the formal summing-up of the whole of the conference proceedings.

He explained that the basic aim of the conference was simple enough—to define the consequences of technology or automation—and the term used was not relevant—and its impact on the individual. He added: "Our basic problem was to distil some sense of direction. We have succeeded well, and we have charted some courses which can have importance for us in our respective countries."

Mr. Kheel continued: "Automation and technology offer us possibilities that this world has never seen in its long history, possibilities for good and possibilities for evil. We wish to be positive—not negative. We do not condemn automation. We would be silly if we tried to; and we seek to realize, particularly for individuals, what its potential truly is. In this process we recognize with regard to employment that technological change has made a significant impact, far greater than from the time the wheel was invented. We think it is a mistake to shut our eyes to the consequences of technological change on the employment structure, if we wish to reap simultaneously the benefits that it can bring."

He recalled that when he helped to form the Foundation in the United States in 1962 and Lord Hirshfield did likewise in creating the British Foundation in 1963, there were many who said that such concern was mistaken, because automation was nothing more than a new form of mechanization; that the definition of automation on a narrow basis indicated that its consequences would be minimal; that to the extent that there was unemployment, principally in the United States at that time, it was not the result of automation but rather of faulty fiscal policies, and that by correcting these fiscal policies unemployment would disappear.

It was true that in the United States there has been a substantial reduction in unemployment in the last 7 years—perhaps by as much as 50 per cent—but it is now evident that in what is regarded as a period of full employment it is not evenly distributed throughout the economy. There are many who still have difficulty in getting into the job market. Yet there has been a lessening of concern about the impact of automation on employment. This was partly due to the present level of employment and partly to the work of the Foundations which had urged that automation should be welcomed.

Sir Denis Barnes had pointed out that in Britain the degree of resistance had lessened, but the patterns of change had speeded up. Mr. Friedrich had observed that technology had affected skilled workers as well as the unskilled. But the impact on the unskilled is predominant, and the impact on those who had not had the opportunity to acquire skills, be they in the ghettos or be they young, is most serious.

Mr. Kheel mentioned that statistics had been presented at the conference to suggest that, based on the experience of the last 10–15 years, 50–60 per cent of the industrial and administrative jobs in our society will be replaced, and many of the replacements, the jobs that will come in the place of those to be outdated, are not known now. He added: "Professor Meshel has pointed out that the ancient names of jobs no longer describe their content, and there are new names for new jobs which only academics can pronounce."

Mr. Kheel is a fervent advocate of collective bargaining, but he insisted that it can be effective only for those workers represented by trade unions. But the question had been raised whether trade unions are an obstacle to the introduction of automation through feather-bedding and other restrictions on work output, and whether the collective bargaining process is so unstable because of strikes as to hinder the advance of automation. He held the view that collective bargaining can make a positive contribution to the introduction of automation.

Frequent mention had also been made at the conference of the desirability of consultation at plants in advance of change, and it had also been pointed out that changes in the employment structure due to automation should not result in the loss of earning capacity of the employee adversely affected. There are problems, however, in maintaining a wage level above the skill requirements of a new job. It had been urged that notification of any lay-offs or redundancies should be given well in advance, and should be reduced to the absolute minimum necessary.

Another tendency of automation had been towards polarization —where the haves have more, and the have-nots get less. There is the prospect of the widening of the chasm between those who can make it economically and those who cannot in a technological society. Other developments could be increases in the tension of confrontation, between the veteran worker and the young worker. And because of the practice of seniority within the unions, black workers tended to develop hostility towards the unions.

Education had been appreciated as one of the most important factors in preparing workers for change. Professor Crossley had stressed that the most important factor influencing the sharing of income is the distribution among the population of human capital in the form of education and training. Professor Kaldor had said that the advance of technology in underdeveloped countries depends on the quality and character of education, and Dr. Kenneth Clark had also placed major stress on education.

There had been differences on the extent to which there should be participation by trade unions in fashioning governmental policy. And there had been useful discussions on the entire question of participation by individuals at plant, trade union, and national level. There had not been any serious question about the democracy of the procedures by which representatives of trade unions are selected or elected, or on the democracies of the unions themselves. Mr. Kheel added: "If anything, I would say that perhaps they are too democratic to the extent that all major decisions, if not many minor, are put to a vote of the rank and file under conditions where the leadership has been presented with obstacles to overcome in persuading the rank and file of the correctness of the position it has taken."

Divergent views had been expressed on participation in management decisions. There was no proof yet that this had significant value for employees. From Britain had come a trade union view that management should manage, and that once labour became part of management the workers would tend to lose their representation.

Mr. Kheel observed: "I personally veer towards the view that trade unions should participate in joint determination or decision-making with regard to wages, hours and working conditions, matters which are well within their competency, and should not assume additional responsibilities, participating in what are purely managerial decisions, even though they would have an effect on labour, because this puts them in a schizophrenic position where they have to think simultaneously what is good for the company and what is good for the workers."

He continued: "Whichever way you take it, the fact remains in particular instances that if leaders have to look towards their membership in order to lead, and if they have to participate in managerial decisions to improve the operations of the company, they are likely to lose out in either one area or the other."

Mr. Kheel declared: "If increasing complexity is taken to be a sign of evolving civilization, then we are becoming a civilized lot. In tracing the significance of some of these complexities, we are probably still at a point of discovery on the meaning and implications of the new technology, but with its growth the opportunity of constructive change is irresistible. In this process we, of unions, management, academies of learning, or other institutions in our democratic societies, have a special and growing initiative in the development of individuals these organizations and institutions are designed to serve. And we have a vested interest in surmounting barriers of thought and tradition, and pushing through frontiers of constructive change. And that cannot be a bad thing."

Mr. Kheel had pointed out the constitution of this type of conference. He explained: "We gather here as private individuals, even though there are government representatives present. We are not here in any representative capacity, even to the extent that we come from labour or management. The views expressed are those of the individuals who spoke them. The Foundations were created in the belief that there was a contribution which private citizens, in positions of significance in our society, could make by joining together and discussing these problems, and reaching conclusions which could be submitted to the decision-making organizations in our respective countries for their consideration."

He concluded: "We have done this in the past, and I think we have, in the process, made a substantial contribution in our respective countries. What has developed at this conference is a validation of the premises on which our two Foundations were created in 1962 and 1963. We had the belief then, and we do now, that automation or technological advance is inevitable, and for that reason alone, since it can't be stopped, even if we wished to do so, we must learn to come to grips with it."

Chapter 2

Is Time on Man's Side?

NOBODY actually asked throughout the conference: How much time has man to prepare himself for adaptation to technological change?

There was no need to do so. Intermittently throughout the week it was clear that the world is already living in an era of change. The "revolution" has begun. It is more obvious in some countries than in others, yet even in the most sophisticated communities the sense of urgency is blunted by preoccupation with seemingly more pressing problems.

Mr. P. L. Siemiller (United States), the international president of the International Association of Machinists and Aerospace Workers, commented shrewdly, "Today in the United States there is very little concern about the displacement of men by computers. There seems to be a common consensus that so long as it isn't raining there is no point in fixing the roof."

He went on: "Although great changes were brought about by the Industrial Revolution over the past 150 years they were minimal as compared to the changes that are being brought about by the computer revolution in a far shorter period of time. It has been noted that for most of man's history human knowledge doubled itself about once every 2000 years. Today, scientists tell us that human knowledge is doubling every 10 years. And nowhere is this explosion of knowledge more apparent than in the application of computers to industrial processes. In 1954 there were only a few hundred primitive first-generation computers in the United States. By 1963 there were 15,000. By 1966 30,000; today, just 3 years later, there are more than 60,000. The number of computers performing jobs traditionally performed by human labour is now doubling every 3 years. And this rate of progression will undoubtedly increase."

Mr. Siemiller added: "Computers are not only multiplying their numbers; they are also expanding their capabilities. During the decade of the 1950's the brunt of computerization and automation in the United States was borne by blue-collar industrial workers. In the steel industry, auto manufacturing, rubber production, oil processing, mining, and every other basic industry, increases in productivity went hand in hand with decreases in job opportunities. Today, as we come to the threshold of the 1970's, it is well known that the future of computerization —and the next displacement of human labour—is coming in offices, banks, insurance, retailing, wholesaling, and the professions."

He concluded: "Although the shape of the future is clear and undisputed by those who know the facts, there is surprisingly little concern in the United States today about the possibility of mass technological displacement."

And Mr. Siemiller went on to explain why so little concern was shown in the United States about the displacement of men by computers; why the substantial drop in unemployment "seems to have nullified the feeling of urgency that existed in the early sixties".

The economic skies in the United States "have been pretty sunny for the past 8 years". "But", added Mr. Siemiller, "there are some rather ominous dark clouds on the national horizon right now. I think that many of those who are complacent today are going to be forced by the facts of economic life to realize that the problem of technological unemployment has not been solved. It has only been delayed. Ultimately and inevitably automation, like overpopulation, is a problem with which societies everywhere will have to come to grips."

And the delay of realization in the United States is explained by what Mr. Siemiller described as "the timely intervention of a series of unusual stimulants".

The first is the war in Vietnam. This war has not only taken millions of young men out of the work-force for military service but it has created unusually heavy manpower demands upon industry of all kinds.

Mr. Siemiller declared: "Automation is not yet a serious problem only because we are spending 80 billion dollars a year—43 per cent of our total national budget—for military purposes. In addition we are spending 5 billion dollars a year to put a man on the moon. Together these expenditures for our defence and space efforts generate jobs for more than one out of every ten workers in the United States."

"These two factors", he continued, "have created enough jobs to soak up the immediate impact of computerization. But other factors have also helped. These include, firstly, the post-war baby boom of the 1940's which has resulted in a far higher than normal rate of new family formations and household consumer expenditures in the 1960's; secondly a highly favourable (but now rapidly diminishing) post-war balance of trade between ourselves and other nations; and, thirdly, the initiation of a so-called war on poverty at home. Although this programme has never been fully or adequately funded, it has helped to reduce unemployment by putting as many as 200,000 hardcore unemployed youths into training programmes. Another factor has helped to delay and soften automation's impact. Though seldom mentioned or even admitted, it is nonetheless, real. And this is the fact that in times of booming prosperity and record-breaking profits many corporations tend to become overstaffed. So long as the economy is going full-blast there is no real need to cut back that part of the work-force that is essentially non-productive. When a drastic economic turn-down comes—and there are now signs of such a turn-down in the US economy—a lot of large corporations are going to start analysing their operations and weeding out all of the jobs that are not absolutely productive or essential. The search for greater economy—the need to reduce labour costs—will also stimulate more computerization. The evidence is conclusive. The problem of technological unemployment has not been solved because it has not really been faced. Recent record-breaking levels of economic growth are not normal. They are the result of a number of unique factors, none of which can be maintained indefinitely. But the opposite is true of technological change. Not only will present rates of technological change be maintained but once the economy tightens they will be accelerated. The paradox we face is this. In the future there will be more people coming into the labour market looking for jobs. At the same time there will be fewer jobs for *people* because there will be more jobs for *machines*. In saying this I am not speculating on future possibilities. I am describing present facts."

But prosperity and increased efficiency have grown together in the United States. This was one of the points made by Mr. Elmer T. Klassen, who is now the Deputy Postmaster-General of the United States, but was formerly president of the powerful American Can Company which he joined as a £5 a week office boy.

He recalled that he had been a corporate employee for more than 40 years as Mr. Siemiller had been a member of his union for 40 years, and during the last 20 years he had represented his company in negotiations with Mr. Siemiller's union.

Mr. Klassen continued: "Now in these past 20 years my company made very substantial technological changes, and during the 20 years that I speak of we had no strike with this union. Now, the next 20 years, I am sure, will bring even greater technological changes, and I know that Mr. Siemiller and his union expects this company to move forward. And if industry fails to move forward everyone loses—the unions, the employees, the customers, the public, the government."

He compared the growth of his company and Mr. Siemiller's union during the past 40 years. He said: "In 1928 Mr. Siemiller's union had 70,000 members and now it has more than a million. In 1928 the machinists rate in the United States was between 50 and 60 cents per hour; 40 years later a machinist in the United States earns in excess of 4 dollars an hour. In contrast, my company's sales in 1928 were 100,000,000 dollars; in 1968 they were in excess of 1,500,000,000 dollars. In 1928 we had 16,000 employees; in 1968 we had 56,000 employees."

Mr. Klassen added: "Our standard of living is the greatest in the world, and this has been helped very substantially by our improved technology. And, while our cost of living has risen substantially during the past 20 years, I assure you that it would be much higher had we not had a continuing campaign in American industry to improve our technology."

Lord Peddie, the deputy-chairman of the Prices and Incomes Board, questioned the pessimistic note of Mr. Siemiller's paper. "Is he suggesting", Lord Peddie asked, "that if automation comes along, greater mechanism and computerization, it means automatically there will be less claim on labour?"

Mr. Siemiller replied "In any country in the world there are only so many jobs, regardless of the number of human beings that populate that country; that's a known fact. Unless we create new jobs and we use our creative spirit to provide full employment, it just doesn't exist. And that is what the paper intended to say was the responsibility—that every human being who wants a job must have one provided. And I said that it is the duty and responsibility of government, in the last resource, to provide jobs for human beings if there are no jobs in that country. There are only so many jobs unless you create jobs, and it is the duty of society to create the job for the man."

Professor Ben Roberts, of the London School of Economics, observed: "I frequently find myself extremely surprised on listening to American speakers on problems relating to their own economy. Here we have the most successful economy in the world in terms of its own technical performance. And yet, at the same time, whenever I listen to an American speaker I hear the most pessimistic account of their economy. I hear speakers say they believe in progress but at the same time express the most deep and horrendous fears about the future. If you take the question of employment which Mr. Siemiller has dealt with this morning, I would like to ask him if he is aware of the fact that his own economy has been creating new jobs at a net rate of not fewer than

1 million a year every year over the last 20 years, so that there are 20 million more workpeople employed today in the United States than there were 20 years ago. I'd like to add that if one makes a calculation, admittedly it is a rough one, the world as a whole has created something like 500 million new jobs net in the last 40 years. This seems to be a fantastic performance in terms of the ability of modern economics, to create jobs as distinct from destroying jobs, which they do as technology advances. And if one is putting into the balance the two things, the destruction of jobs and the creation of jobs, the net situation at the present time is a highly favourable one, and I see no evidence to believe that we are likely to face a situation in the foreseeable future where the world will face, in fact, the kind of pessimistic situation which you have indicated, nor do I see that situation in the United States."

Nevertheless, Professor Roberts realized that there are pressing problems in the United States. These problems also affect many other advanced industrial countries, including Britain, though perhaps not to the same extent as the United States. He felt that collective bargaining alone might not have the dynamic capacities to meet these challenges, and it is only in a newly developing tripartite system where governments have to play a major role with their appropriate agencies that these problems can be met.

Mr. Siemiller explained that he did not claim that collective bargaining could produce the jobs required, but it did protect jobs in existence and it protected "human beings from being removed from their jobs because of technology without something being provided for the individual for the future".

But he insisted that "with all of the progress of the world, we don't have enough jobs in the world. If we did, we wouldn't have the famine that we have in India and other countries where there are available work forces that could work. Their jobs are needed. But although America is one of the most highly industrialized countries of the world, we still need jobs, and we need the creative spirit of management, of labour, and of government, to create the necessary jobs so that every human being who wants to work, to earn a living, will have the opportunity to do so."

Mr. Theodore Kheel, the leader of the American delegation, demurred at Professor Roberts's estimate. He observed: "Professor Roberts has grossly underestimated the productive capacity of the United States to produce new jobs. We are producing not only a million new jobs a year—we're producing closer to 4 million new jobs a year." But, Mr. Kheel added, "We are also automating out of existence approximately the same number. We have been able to maintain a kind of balance primarily because of the various stimulants to our economy which were set forth by Mr. Siemiller." [Obviously, Mr. Kheel had not appreciated Professor Roberts's point of a *net* increase in jobs.]

But in Britain, too, a significant change has taken place, and this is a continuing process.

Sir Denis Barnes, Permanent Under-Secretary of State, Department of Employment and Productivity, in a succinct survey of changing job content in Britain, revealed that in the last 50 years the working population has risen from just over 18 million in 1911 to more than 24 million in 1966.

The numbers employed in manufacturing industries have risen by about 2¼ million to a present level of nearly 8½ million. But this is a slight increase only in the percentage of the working population employed in manufacturing to a figure of 35 per cent. There has been a very big decrease in those employed in the primary industries, agriculture, and coal-mining, both in numbers and proportions. In 1911 agriculture employed about 1½ million and mining and quarrying about 700,000—together about 15 per cent of the working population. Agriculture now employs about 400,000, mining and quarrying half a million—together less than 5 per cent of the working population.

There has been a substantial increase in the transport and service sectors, which now employ more than 50 per cent of the working population. During the last decade alone, the numbers here have increased by nearly 50 per cent. Transport, after increasing to a peak in 1951, now seems to be declining. Since

1951 there has been a considerable shift from rail to road transport, and, of course, an increase in the use of private cars.

Employment in building and contracting and public utilities has increased steadily. They employed just over a million workers in 1911—nearly 6 per cent of the working population—but they now employ more than 2¼ million or nearly 10 per cent.

It seems likely that these trends will continue and that there will be a further decline in the primary industries offset by an increase in the services sector.

Within the manufacturing industries there has been considerable change. In 1911 the metal (and metal using) industries and textile industries were about the same size, each employing nearly 2 million workers. Employment in the metal and metal-using industries is now well over 4 million. Textiles now employ fewer than three-quarters of a million. Employment in the paper and printing industry has doubled. While the numbers employed in food, drink and tobacco have increased, they still count for about 3 per cent of the population—the same as in 1911.

Skilled workers now represent a rather higher proportion of manual workers than in 1951, semi-skilled and unskilled a rather lower proportion. The only group of manual workers which represents a higher proportion of working population than in 1911 are skilled engineering workers. Most of the increase in this group took place in the manufacturing industries, although an increasing number are now being employed in the services industries.

The proportion of administrative, clerical, and technical workers has been increasing rapidly since 1911. Clerical and allied occupations account for the greater part of the growth. They now account for about 14 per cent of the working population, and there has also been a rapid expansion in the professional and technical occupations, which now represent about 10 per cent of the working population.

Sir Denis wrote: "These changes in occupational structure are similar to those which have occurred in other developed countries, particularly the relative decline of manual and relative growth of 'white-collar' occupations. In all these countries they are the result of changes in technology, which reduce the need for manual work; an increase in the size and complexity of industry, which increases the need for managers, administrators, and clerical workers; an extension in the provision of public services and therefore of government activities; and a higher demand for the services provided by the private sector as standards of life improve."

He gave examples of the technological innovations which are most clearly apparent to individuals at work. There are the use of electronic computers, developments in instrumentation and automatic control, the numerical control of machine tools, advances in communication technology, improvements in machinery, the mechanization of material handling, new developments in metal processing, technological advances in agriculture and transportation of all types, progress in power production, and the continuous development of new products and materials.

Sir Denis added: "The most significant technological advances have taken place in the chemical, oil and steel industries, transport, commerce, and communications. In Britain, the move towards automation has been more rapid in office work than in manufacturing processes, but wherever it has occurred it has usually called for extra skills, for example, those of electronics experts and computer programmers, and reduced demand for unskilled workers previously engaged on repetitive work or heavy manual jobs. Semi-skilled workers are still required but with the emphasis on responsibility, alertness and understanding rather than manual capabilities."

Some interesting examples of the effect of technological changes were produced by Sir Denis. On the railways, for instance, these changes have effected "some very dramatic manpower savings". The changeover from steam to diesel resulted in a decrease in numbers, partly because firemen were no longer required and partly because diesel engines could be used much more intensively. Similarly, mechanized track maintenance including the use of automatic tamping and lining machines requires a considerably smaller but much more highly trained work force. Automatic signalling has enabled a much larger number of signals and train movements to come under the control of a single power signal box. To take one particular example—a single power box installed 3 miles from Paddington will control all train movements into the station and will ultimately replace fourteen manually operated signal cabins.

In iron and steel, too, technological changes have effected substantial manpower savings. For example, the introduction of electric arc furnaces and the use of electronic control systems allow production to be achieved with a very much smaller labour force. In one company as a result of such changes the numbers employed in actual steel making dropped from 1900 to 600.

And towards the close of the conference, Mr. Dai Davies, the leader of the British steelmen's trade union, gave a vivid example of the effects of change on Britain's tinplate industry. He recalled that in 1938 tinplate production in Britain was spread over 300 mills with a labour force of 27,000 producing 20 million boxes of tinplate a year. During the last 16 or 17 years tinplate production in Great Britain has been concentrated into three mills with a labour force of 3000 producing 30 million boxes of tinplate a year. Mr. Davies added: "Both figures in themselves show that we have in fact passed through a technological revolution. And I think the important point about it is that the technological revolution took place with comparatively few difficulties at all. It was carried through because there was the fullest co-operation between the trade unions and the employers."

But one of the most intriguing features of Sir Denis Barnes's contribution was his assessment of the reactions to change on the part of workers, employers, and governments.

He believes "The personal reactions of employees to change, cannot—except at unthinkable economic and social cost—prevent technological changes, but they can certainly influence the pace and profitability of innovation. They will certainly influence relations between employees and managements when dealing with all issues which arise between them, not only those just related to the changes. They will influence the general tone of an industrial society."

Sir Denis recognizes that the reaction of the majority of work-people, and some of the management, to technological changes is primarily anxiety or fear.

He added: "There will be some, of course, to whom the changes will represent opportunity but these will be a minority. Amongst the majority some will be anxious that they will lose a job and be unemployed; others may not fear unemployment but will resent the need to change employment at a time, and perhaps on terms, imposed on them and with little choice; others who stay on in the same employment may find their job changed and have doubts about their capabilities; others will fear the effects on their status or their prospects of promotion. These are immediate and serious problems for each individual."

In any assessment of the factors which influence these attitudes the general economic and social environment is clearly important. If a country or industry is prosperous and there is a reasonable expectation of employment security, in a general sense, attitudes to change will be more favourable. But, Sir Denis asked, "How can we assess whether unemployment is at a level and of a character which does not cause too great uncertainty, and what do we mean by employment security? If society's conception of these is unrealistic, then the probability will be that its attitude towards the technological change also will be unrealistic. It can be argued that the major and continuing rate of change in the pattern of industry and occupations to which I have referred, have such implications for mobility and the need for training and retraining that they must be an important factor determining the economic and acceptable level of unemployment. It may be that an increasing proportion of unemployed will be necessary to provide the flexibility essential to meet changing manpower needs. It is difficult to persuade people of this, but important that it should be accepted."

He continued: "What do we mean by employment security? Certainly the concept must be widened beyond the traditional resistance to redundancy and preference for work sharing to preserve a particular job. It should take account of the provision made by society and individual managements to enable workers to change their jobs and employment with minimum loss of income or other benefits and broadly be based on an assessment of employment prospects for an individual from youth to retirement."

And Sir Denis reached this interesting conclusion: "In Great Britain the attitude of workpeople to these problems has changed in recent years. While the shifts in the pattern of employment due to technological innovation have speeded up, the degree of resistance has diminished and in my view is less than we experienced, say, 10 years ago. But we are still a long way from the degree of acceptance and positive co-operation which are needed. We shall only achieve this if the policies of managements, trade unions and governments are right."

And what of managements? Sir Denis argues: "The immediate responsibility for introducing technological change with a minimum of anxiety for and maximum acceptance by employees falls unavoidably to management. Of course, management can only do this as part of the process of pursuing its general objectives of efficiency and profit. But the consequences of all managerial decisions for the employees of an enterprise should be considered and this means taking account of the reactions I have described. Manpower problems, manpower planning, should be given the same attention as investment, marketing, technology, and other production problems. Technological and organizational changes to improve efficiency are not governed by fixed laws or by a fixed time-table. There may be a range of alternative solutions which will have more or less favourable consequences of technological change which should be drawn up in good time. Managements should be organized so that those concerned with manpower participate at an early stage in planning the introduction of new processes. If altered manpower requirements are assessed in sufficient time, a plan can be made that coincides, as far as possible, with normal wastage, periods of business expansion, and the capacity of the firm to retrain or transfer existing employees. Employee representatives should be brought in not simply to be informed but to secure their participation in the solution of the manpower problems. Managements have probably more than they realize to gain by discussion and negotiation with their employees and their representatives on the detail of manpower plans. This process can not only reduce the adverse effects of change on employees but increase the efficiency of the change itself and therefore the future prosperity of the firm."

Sir Denis thinks that many managements need expert advice on manpower planning, their organizational structure, wages systems, methods of employee participation and negotiation, the handling of redundancies, and training. It is reasonable to expect them to look to employers' organizations for this advice.

And he concludes that "in recent years an increasing number of managements have developed policies which follow the basic principles I have mentioned. Some employers' organizations are developing their functions on the lines I have indicated. The need is for these approaches to become more general throughout industry."

And as for government, Sir Denis observed: "The objective of government should be to reconcile the effects of technological and structural change in industry with established expectations of employment and income security, job satisfaction, and work adjustment. If government action is inadequate, workers will resist change regardless of the changes in attitudes and policies of managements and trade unions. Effective government policies are necessary for economic expansion, industrial efficiency, and social well-being."

Sir Denis summarized such existing government legislation as the Contracts of Employment Act, 1963, which gives both employers and employees minimum periods of notice to terminate a contract of employment, and the Redundancy Payments Act 1965 under which compensation is paid to workers who become redundant after a minimum service of 2 years.

He also explained that the public employment service is being developed. Its objective is in effect to create a national manpower agency which will deal more efficiently with the employment problems of changes, with fuller knowledge of the labour market, more advance information about likely changes, more expertise on occupational guidance, and a more efficient regional and area organization.

The entire system of training and retraining is being reorganized and rationalized under the Industrial Training Act 1964. A training board has been set up for each major industry.

The activity of these boards is financed by a levy on employers which they can recover in whole or part by obtaining grants from the boards upon proof that they give satisfactory training. This includes all forms of training for all ages and levels in eachindustry. The duty of the boards is to ensure that sufficient training is provided. Each is required to make recommendations to the government on the nature, content and length of training for occupations in its industry and also on further training. They have the power to compel employers to make returns, prepare estimates of skilled manpower requirements in the future, and draw up plans.

Since 1963 the number of government training centres has increased from 13 with 2600 training places to more than 40 with approximately 9000 places and the programme of expansion is continuing.

On the face of it, therefore, Britain seemed commendably prepared for dealing with the dislocations and upheavals created by rapidly expanding technology, but there was no smugness about Sir Denis Barnes's survey. In fact, he foresees tremendous social changes before the end of this century, well beyond the scope of existing plans.

He declared: "I would not claim that these policies were evolved with foresight and after study as a far-sighted plan to deal with the problems discussed in this paper. Most were introduced under the pressure of particular problems. I think the same is true of the changing attitude and policies of managements and trade unions. This does not make the measures and policies less valuable."

Significantly, Sir Denis adds, "It does, however, lead me to suggest that though government, trade unions, and managements are reacting to the problems of the present and immediate future, they are still not giving enough thought to the longer term."

Still unresolved, therefore, is the problem of how much time is left before further technological developments make a more dramatic impact than ever before on established procedures. Britain has the advantage of an established tripartite system in which representatives of government, managements, and workers are accustomed to meeting each other to formulate plans to meet any problem.

But, as Mr. Patrick Fisher, of the TUC, observed in a slightly different context during conference, "One difficulty we are running into is the problem of the time scale. I think that if you make any kind of prophecy, and if you give yourself enough time, judging from the science fiction authors of the past, there's a fair chance that those prophecies will come to fruition some time in the future. One of our difficulties is to say when any problem is to require action. Do we have 5 years, do we have 10 years, do we have 15 years in our particular countries to meet particular problems? There might be a danger", mused Mr. Fisher, "that one tends to get to the stage of frightening oneself."

About some of the Israeli spokesmen, however, there was a sense of philosophic realism, the more impressive because of the exceptionally difficult problems facing the newest of the industrial countries represented at this conference.

Thus Mr. Ben-Ami Friedrich refused to be flustered. Professor Jay Tabb, when introducing him, explained that Mr. Friedrich is an industrialist who manages a modern plant with modern equipment. Professor Tabb added: "This modern equipment is operated by a work force, the majority of whom prior to coming to Israel had never seen an industrial plant. Mr. Friedrich, from his own experience, knows that he succeeded in training an industrial work force to operate an efficient plant. That's probably the reason why he's not too excited about the problems that will stem from technological change. In either case, while he recognizes that there will be change in the occupational structure of the work force, the equilibrium will be reached in the long run, and we generally have from 3 to 5 years in which the work force can be trained."

It was against the broad historical sweep of recorded industrial history, from the discovery of fire by Ug through the alarms of the Industrial Revolution of 150 years ago, that Mr. Friedrich asserted that there was nothing particularly new in having to face problems. He agreed "Automation has brought about changes in the stratification of the labour structure. It has affected the image and the demands of work. Even greater than its impact on the present, have been the theories and hopes it has aroused for the future. Has the effect of automation been so great as to confront the responsible powers in the economy with the need to take extraordinary measures in order to moderate or advance its future? The idea of automation has so powerfully affected, even prejudiced, the mind of modern man that the interpretations are insufficiently reliable. In trade unions and management, in government and academic circles, interpretations based on regional differences or on certain branches of industry only have been responsible for this unreliability. Prejudice rather than judgement does not constitute a well-founded basis for predicting future developments and making the necessary decisions."

Mr. Friedrich added: "Many of the hopes and much of the restlessness associated with automation and its impact on society are the outcome of the public debate about the immediate future and what is new in it. The alarm, the noise and the prejudice for and against should be firmly set aside. The transition from one industrial level to another and all further technical advance is essentially a continuous process. Each of the advances may be described by a curve which changes according to the various parameters. This process is accompanied by unpredictable overlaps and sometimes is marked by instances of sudden jumps, occasioned by unforeseen economic developments or disturbances such as war. One other point should be kept in mind here. It is an assumption, that on the basis of the technological development of the last 10 to 15 years conclusions may be drawn as to the next 10 or 15 years. If this assumption is correct it follows that 50–60 per cent of the industrial and administrative occupations in existence today will be replaced by others, some of which are not even yet known. However, even if this happens, it will still be in only very few cases a process of development by jumps. There will be, in any case, a long drawn out process and the possibility of selective disengagement from the specializations known and required today. This disengagement is not conditional on economic development solely. Technological obsolescence influences most the equipment and machinery requiring the highest rate of amortization. This process, though, contributes to changing professional knowledge and during the 3–7 years that this takes place, we can prepare the new cadres. The tremendous capital investment made specifically in heavy industries during the past 10–15 years will not be scrapped suddenly when it is financially amortized. Small current changes will also contribute substantially to the overall continuous change. All of these changes will significantly influence the overall economy only if a great number of individual plants are able, or forced, to introduce automation. It will lead to changes in the national economic social life only in so far as they can be assimilated into the current economy."

Mr. Friedrich continued: "The setting up of entirely new industries, based on developments as yet unknown, or still technically unripe, will require a period of at least another 3–5 years, in addition to the 5 already mentioned. In many cases, of course, the time required will be even more. We may consider, then, that we have 10 or 15 years at our disposal to accommodate ourselves to such changes as may be imagined from where we stand today. During that time it may be assumed that the demographic changes that take place with no relation to automation will themselves provide at least a partial solution to the problem."

He added: "I do not believe that the importance of the subject has been diminished by removing some of the unreal problems which have become attached to it. It is necessary to keep the problem in proportion, otherwise we should judge every technological or cultural development in terms of its imaginary dangers or shortcomings, and consequently our conclusions would be both blinkered and distorted."

Nevertheless, Mr. Yosef Almogi, Israel's Minister of Labour in opening the conference, had explained that his country, having to devote 20 per cent of its gross national product to security, could not "afford the luxury of just stepping slowly towards raising technological standards, but on the contrary we have to rush towards these changes". And Mr. Aharon Becker, the leader of the Federation of Labour of Israel, while acknowledging that no serious trade union would "regress to the time of the industrial revolution and nobody will try to turn back the wheel of progress", added that while automation needs to be introduced and accepted, it should "not be accepted automatically".

Undoubtedly, there is a variety of attitudes towards the timetable of massive changes. There is a wider consensus of agreement on the probability that involvement with technological change, and its consequences, both industrial and social, will vary considerably from country to country. It is probable that the United States, the more advanced countries of Western Europe, and Japan, will have to deal with the "new revolution" long before most other countries, and it will be the more sophisticated firms in the industrial countries which will set the pace.

But the awareness that time is not necessarily on the side of the cautious or the timid was almost electrically stimulated without warning in the very closing session of the Jerusalem conference.

One of the innovations of this conference was the introduction of an international "brains trust", a representative panel of academics, management, and trade union representatives. There were no prepared papers and no foreknowledge of the questions to be put by the conference. In fact, Mr. Leslie Blakeman, a former Ford Company executive recently appointed to Britain's new Commission on Industrial Relations, who was the session chairman, obligingly recapitulated the main issues and submissions of the whole conference as a stimulus to prospective questioners.

With disarming tentativeness, Mr. Cyril Plant, the leader of Britain's Inland Revenue Staff Federation and a member of the TUC general council, asked Mr. B. John Hargreaves, an executive of IBM United Kingdom Ltd., "I wonder if Mr. Hargreaves could engage in some crystal gazing as to what might be the development of technological change especially in the computer and automation field over the next few years, and what he feels might be the need for the creation of more jobs to absorb any labour displaced?"

Extemporaneously and confidently, Mr. Hargreaves replied: "I think on the technical side the big switch in objective has come from trying to keep pace with paper to the new idea of eliminating paper because paper is an inefficient means of communication—anyhow in business life."

He continued: "If one looks ahead one can put a date on some of the changes which will take place. For example, cards and paper tape in the computer world will become obsolete by about 1975. The spoken word, or what we call oral input in the computers, will be general by about 1980. Books and libraries as sources of factual material will be superseded by computer files by the mid-1980's. By the 1990's we shall be building houses with electronic walls which will allow people to receive information from computer networks in their homes and in their offices. By the mid-1980's we can expect that doctors, and perhaps other professionals, will have terminals as part of their equipment, so that a doctor can plug into a computer network for diagnosis and things like that. By the beginning of the 1980's we shall experience one-million word desk size or possibly even briefcase size computers, using laser memories. We can see chemical memories and chemical storage in computers in the next 20 years."

Mr. Hargreaves added: "Perhaps it is relieving to know that the cost of all this is coming down and that by the 1990's the cost of computing will have decreased by a factor of 100."

In answering the second part of the question, dealing with the effects of more advanced technology, Mr. Hargreaves revived other problems which had engaged the conference during the week.

"Immediately", he continued, "the price to be paid for continuity of work in a changing world is constant relearning and the preservation of an agile mind. Perhaps the battles of the union in this decade should be the right to transferable pensions, the right to retraining and so on."

He declared: "We see the problems of a divided world, which has been caused largely by technology, and yet the problems of which can be solved by this same technology. Certainly a single world was once morally desirable; today it is politically and socially vital."

"Looking ahead", Mr. Hargreaves surmised, "I suppose the biggest thing that is happening is the move from a manufacturing to a distribution society. Already we see signs of this distribution-service economy. And this means not an idleness or a ceasing from work, but a released capacity which will allow man to do other things. And this means we have to train him to do other things."

He concluded: "I suppose, finally, the sort of challenge that faces us is that one can see by the last decade of this century considerable control through technology over our environment. One sees the obvious possibilities of destroying our environment as well. Therefore the release that the machine provides to man, the released capacity which it gives him, is the capacity to think about what he should be doing when the drudgery and the work unfit for human consumption is removed from him."

Clearly, the blessings of technology are not unmixed. There are so many adjustments to be made, not only by man alone but by communities and nations.

Chapter 3

Never Too Late to Learn

IT WAS not surprising that in his official summing-up at the end of the conference, Mr. Theodore Kheel should claim that if any single concensus emerged from it one stood out well above all others. And that was education—not just the curricula of primary schools, or higher schools, or technical colleges, or even of universities, but the whole vast panorama of instruction in all its forms and in the variety of its benefits.

With succinct clarity, Mr. Kheel summed it up as "learning from the cradle to the grave".

In one way it was a surprising revelation that education and training should have commanded such a pervasive domination, for the main themes of the conference were, after all, the inevitable expansion of technology and man's adjustment to it. But in another way, it was obvious that to divorce from these matters the essential need to produce more sophisticated people for a rapidly changing world would have been fecklessly unrealistic.

Few of the major papers, however superficially contrasting their subjects appeared to be, failed, to a greater or lesser extent, to draw attention to the importance of education and training; yet it was exceptional for any of them to probe deeply into the techniques of instruction, partly from preoccupation with more specific aspects of technology, partly from an assumption that the benefits of improved educational facilities were too self-evident to need elaboration. There were however, some exceptions.

Clearly, what emerged from a review of the whole week's proceedings was the urgent need for an authoritative investigation in depth of the adequacy of existing educational and training facilities. In an era in which millions may be called throughout their working lives to change their callings and their skills possibly two or three times, it becomes increasingly evident that the termination of formal education becomes little more than a preliminary exercise to the acquisition of subsequent mental adaptability.

One of the most intriguing features of the many scattered references to education and training throughout this conference was an impression that there are some significant differences in national attitudes towards these matters. Of course, there can be no dogmatic conclusion on this issue. Participants at this conference were expressing their own personal views, sometimes highly contentious and provocative, and it was obvious in some cases that they were in conflict with academic and educational experts in their own countries.

Nevertheless, one got the impression at the Jerusalem conference that there exist distinctive, strongly held views on this whole vast business of education. It may be challenged as superficial, but there is some evidence for declaring that the British are fairly satisfied that they will find a way of overcoming whatever shortcomings there may be in their educational and training facilities; that the Americans, alarmed by the magnitude of their discernible ethnic and economic problems, are conscious that a massive overhaul is demanded in their educational and training set-up; and the Israelis, alert to the complexity of having to assimilate an "army" of illiterate immigrants, and traditionally absorbed in the fascination of learning, are stimulated by the challenge of effective tuition, the provision of adequate facilities, and the absorption of knowledge.

It is proposed, therefore, in this summary of conference proceedings, to justify these inferences by joining together in "national" groups the contributions made by participants on the spectrum of education. The fact that some of these speeches were made on different days and in the context of discussions on differing topics, serves to emphasize the preoccupation of so many participants with the supreme importance of education.

Undoubtedly, priority must be given in this respect to the American contributors, partly because of some of the disturbing views expressed and partly because of the frightening magnitude of their problems.

The most startling proposals of all came from Dr. Kenneth B. Clark, who is president of the Metropolitan Applied Research Center Inc., and who was described by Dr. James R. Dumpson as "a unique individual on the contemporary American scene".

In his paper, Dr. Clark wrote with the eloquence of despair and the courage of dedication.

He made the dramatic demand that American business should create its own educational system. The Negroes, the White dispossessed, and the drop-outs from society, he claimed, have a clear picture "of the hyprocrisy of our society". The government, through its judicial, executive and legislative branches, has tried to alleviate the social, political, and economic problems inherent in the anachronistic form of American racism. Philanthropy has not worked, either, in preventing a deterioration in the "pathology of American cities".

So Dr. Clark urges American business and industry to commit themselves to social change—not for sentimental reasons, but as a matter of hard business.

He explained: "Intelligently planned programmes for positive social change in American cities and in the nation—and in the world—are consistent with the needs of business and industry, for growth—for survival. They put excessive economic burdens on the business and industrial community. Specifically, they exclude a significant percentage of human beings from the consumer market. The present racial arrangement in America provides an underdeveloped economic liability that is 10 per cent of our society. This is apparent not only in terms of consumption but also in terms of production. Built into an anachronistic racial system, is a clear brake on the utilization of available human labour pools by artificial restriction of the labour market on the irrelevant grounds of ethnicity or colour. If American business and industry continue to accept this, they must be considered an accessory to the limitation of their own efficiency."

Dr. Clark declared: "Business and industry must build into industrially structured training programmes remedial compensations for the inefficiency of the public schools—reading, arithmetic, oral and written communication, the major defaults of American schools. Academic training, job training, and jobs must be part of the total package."

He continued: "Business and industry are being offered shoddy products by our schools and being asked to use them as if they were not. Education is critical. It is my considered opinion that the most significant factor in urban instability in the United States is the 'social dynamite' produced by the criminal inefficiency of the inner city public schools. My own persistent pessimism about the extent of improvement one can expect in the public school system has led me to suggest to American business that they take a leaf out of the history of the church and set up their own schools. Not vocational schools—not specifically industrial training schools—but schools for pre-kindergarten through the twelfth grade to demonstrate that the efficiency criteria which seem to work so effectively in business and industry can and will, when applied to the business of education, produce a more effective product than that now produced by the public schools."

Dr. Clark recalled: "The first firm to which I suggested this, one considered the giant of American business and industry, surprised me by being scared to death. They had been brainwashed by the educators who told them that education was the business of nobody else except the educators. Well, if that is the case, we are really lost."

He continued: "I suspect that the significant breakdown in the efficiency of American public education came not primarily from flagrant racial bigotry and the deliberate desire to create casualties, but from good intentions, namely the sloppy, sentimentalistic good intentions of educators to reduce standards in the education of low income and minority group youngsters, to leave them in a state of amorphousness and, thereby, on the grounds that teachers should not demand of these children what they demand of suburban children, make it possible for non-education to be alibied. I urgently warn business and industry not to follow this pattern of sentimentalism under the guise of doing good, under the guise of compensating for guilt and past injustices. Ghetto youth will accept only evidence of respect for their humanity, their capacity as human beings, their ability to learn the way other people learn, their ability to perform with the same standards by which other people's performances are evaluated. And in respecting these, business and industry will be giving not verbal, but genuine, acceptance of them as human beings."

Dr. Clark concluded: "All human beings must be accepted in terms of their humanity and not rejected in terms of the irrelevancies of race or class or sex or religion or geographical accident of national boundary. What is at stake is nothing less than human survival. If the world deserves to survive it will mobilize the intellectual, the economic, the practical planning resources which will make survival possible. This will be the ultimate and practical morality."

Dr. David G. Moore, the Dean of the New York State School of Industrial and Labour Relations, admitted that America's potential for economic growth in the future may well depend on its ability to use the 20,000,000 Blacks which are its "last available source of manpower". He realized that in his "desperate reliance on business and industry, Dr. Clark appears to subscribe to the unsentimental pragmatic objectivity of American business, its basic ability to take a problem, technical and, he hopes, otherwise, and deal with it simply as an issue that needs resolution or as an obstacle to be overcome."

Dr. Moore observed: "Dr. Clark is very much aware of the conflict in values which currently pervades the American scene. He has placed himself squarely on the side of all the virtues, and he sees American business and industry as best equipped to inculcate these values in ghetto youth. He doesn't say so, but I suspect that he fears political solutions which could result in a perpetuation of cultural poverty through make-work jobs, welfare, and hand-outs."

Dr. Moore concluded: "Whether or not American business is up to the whole task is questionable."

And Dr. James R. Dumpson, the Dean of Fordham University School of Social Service, also had his doubts about Dr. Clark's ideas.

Dr. Dumpson commented: "Dr. Clark properly identifies basic education as critical, but I must join Dr. Moore in rejecting the proposition that business should enlarge its role to include basic education. I share Dr. Clark's dissatisfaction with American public education, particularly as it has demonstrated its incompetence to prepare minority youth for living. But I make a distinction between fundamental education, preparation for total life, and training for jobs. Minority youth must live in worlds other than the work world, and what is needed is an appropriate input into the educational processes of the characteristics of modern business that Dr. Clark identifies."

He continued: "In addition, a reorganization of basic education is wanted in the hands of reoriented educators to the needs of Black, Puerto Rican, and other minority groups, including disadvantaged Whites. This can and must be achieved not by giving to business, with its middle-class profit-orientated values, the responsibility of basic education, but by giving to parents of Blacks and Puerto Ricans and other minority groups control of their educational system, with educators at the helm, through which the children must move. Training for employment is another matter, and I am quite prepared to support Dr. Clark's prescription for the role of business and industry in this area. But while a reorientation in the education of ghetto youth is essential if they are to be motivated for jobs that technology will require, that reorientation must include an element not frequently mentioned; I refer to preparation for change."

Dr. Dumpson explained: "Little that we train for today or tomorrow is likely to be useful in the individual ghetto youth life of the future, unless it includes the readiness and the capacity to change his vocational role probably two, three, or four times during his working life."

Mr. Vernon Jirikowic, director of research of the International Association of Machinists and Aerospace Workers, thought that entrusting American business with the job of basic education was "quite risky".

He added: "If there is any indication in our ability to convince American business to establish apprenticeship programmes, if that is any indication of their willingness to extend themselves further in the field of education, I think Dr. Clark is quite unrealistic. We as an organization have been trying for many years, along with other labour organizations, to broaden apprenticeship programmes as such, in addition to upgrading journeymen programmes. And, notwithstanding our efforts in these directions, we can't claim a lot of success at all."

Mr. Jirikowic concluded: "I think that the building trades have done a tremendous job in recent years in bringing underprivileged youth into these programmes. Primarily through some of our unions, underprivileged youths have been tutored to qualify for apprenticeship entrance exams. So when Dr. Clark makes generalizations about labour organizations not being sensitive towards the needs of training our youth, I simply must have to say that I don't think he knows what the facts are. We, along with other organizations, really believe that where one earns a living one should also learn, not in a casual manner but in a systematic approach towards developing his full potential. Such programmes, if jointly administered by labour and management, can contribute immensely to meeting this very severe problem."

Miss Barbara Gray, of the American Foundation on Automation and Employment, acknowledged that the people who got out of the educational system, who were not able to finish any kind of basic education, were a very sizeable number. She added: "Being in school for 12 years doesn't necessarily mean that you read or write, and that's what Dr. Clark is talking about. When our Army finds that they have to reject 33 per cent of our young people who cannot pass the armed forces qualification test, which is basically an aptitude test, it means that our educational system has failed. I think what Dr. Clark was pleading for is something to make it change, and that's what all of us, who have to do with the have-nots, are pleading for. Paul Moore of Columbia University made a very significant study on change in education in the United States, and he showed that it took 40 years for a well-established principle to take effect generally in the United States. I talked to him a few days before he died, and he said, 'At the present time [and this was only a few years ago] that has been reduced to 25 years'."

Miss Gray continued: "In an automated age we cannot wait for this pattern of change to improve, and I think what Dr. Clark pleads for is a change which will respond to pressure from business and industry, and that to me includes management and labour. Both have run magnificent experiments in education when their needs called for a particular pattern. I am an adviser to several industries. I am watching one industry with a programme going in Harlem. Youngsters who were at the bottom of the educational barrel are being turned out by the kind of motivation that comes when this big industry takes them under its wing and provides a course and a ladder from which they can get in and out to a worthwhile job. It is practically a miracle."

She concluded: "Youth have it in Israel. I wish we could all see the kind of school I visited this noon, where the rejects are considered by the Israelis as still salvageable. Reject Zero seems to be, to me, their goal. I think that is what, hopefully, could come if industry would put the same effort into bringing about educational change as it does for its own profit and its own production."

Although Dr. Isador Lubin, economic adviser to the Twentieth Century Fund, was dealing primarily with the plight of the dispossessed in the United States, he soon came to the importance of education.

He declared: "The emphasis must be on education. It is the first line of attack that must be expanded so as to enable people to have the skills and the know-how that goes with employment in our modern technological system. Any advance in our already technologically advanced society must be based on mass education, and mass education in the United States means hundreds of thousands, possibly at least a million, more people engaged in teaching and in running our educational institutions. There are a million jobs right there, ready to be filled by those who are given the necessary education and training."

Dr. Lubin referred to the American Government's decision to control inflation. There has been a cut-back on government expenditure. He asked: "What does this mean? It means that the training of those people we felt to be so important as to bring them into industry is going to be cut down. Already the effects of the cutting down on the Federal budget are being felt in the activities of the states, who get federal aid for their activities. In the City of New York the payments for welfare are to be cut and our expenditures on education are going down. As a matter of fact, the president of one of our great institutions of higher education in New York has tendered his resignation, saying that with the budget committed him for next year he cannot take in any new students next September because he just will not have the money to take care of them. Day centres for children are going to be cut, and one could go through the list. One of these projects which has been very dear to my heart is the system of Summer Youth Programmes, in which youth who are out of school during the summer months can get employment and in the process of getting employment not only get an income, but learn skills, learn a trade, get a feel of what is involved in the economic progressive policy."

Dr. Lubin reported a serious short-fall in the training of professionals projected for the middle 1970's by the Presidential Commission in 1960. Among the examples he quoted were: "In the field of health, between now and 1975 we ought to be producing approximately 8000 new doctors every year, but according to the present set-up we are only going to produce 750: in dentistry we need close on 50,000 new dentists every year, but up to now we have never produced more than 3200 a year; and we need

20,000 new social workers every year, but we have never produced more than 5000."

Mr. Theodore Kheel, in commenting in another discussion the following day, wondered whether the inevitable changes following increased technology did not call for "some dramatic changes in our educational system". He was not referring specifically to the United States but considering the universal preparedness of educationists in many parts of the world. He agreed with another speaker that "we seem to be most traditional and most conservative when it comes to the young". He mused: "And are we not faced with a crisis of major proportions with regard to our educational system, and with regard to vocational training, and are we not required at this point to say something more merely than that the answer is education and vocational training?"

He returned to the subject in the closing session of the conference when he observed: "I don't think there is any lack of interest on the part of labour or industry in the importance of education and job training."

Mr. Kheel continued: "What is new is the suggestion made at this conference that industry, and I would add labour, should consider, as Dr. Clark proposed, as a demonstration of what industrial efficiency means, to show the public schools how it should be done. I question the validity of this. I think it is important to have academic training as part of job training, where academic training has been deficient. It is important to encourage improvements in our public school system, but I believe it would be a mistake for private enterprise to enter into the field of public education. Be that as it may, it does illustrate our concern with the importance of education. And it does pose the question of where should the private sector end dealing with job impact, and the public sector begin?"

Mrs. Mary Conway Kohler, director of the National Commission for Resources for Youth, recognized that the conference was concerned about the dispossessed because, with technological change: "we also have need for not necessarily greater skill but for greater education to be able to have that agile mind to be able to change."

And then, Mrs. Kohler added wrily, "And still our educational institutions are dominated by educators!"

She recalled a former Commissioner of Education, who said, "You cannot leave education to the educators, it is much too important." Mrs. Kohler continued: "I think that what Kenneth Clark was trying to say when he said, 'We have to get the consumer, the person who is producing this technological change, to make demands on the educational system and to experiment for them and show them the way for change' is the same thing."

"In each country", she added, "we simply have to find new methods of education which, quite frankly, now the educators cannot give us."

Mrs. Kohler concluded: "Education still, in most countries, is a screening process to screen out the people that were not suitable for education when we had a way to employ those people, when unskilled labour was still needed. And I don't think we have any answer as to how to accelerate it, unless we get some outsiders in to promote a different kind of approach to education than we have now."

Israeli appreciation of the importance of education may not be very different from American, but its experience is. Despite their own peculiarly complex problems, the Israelis have relished the advantages of creating new institutions, untrammelled by the weight of established academic formalities. They adapted what were considered the most effective methods of European and American universities and higher educational institutions, and were fortified by an enthusiastic thirst for knowledge shared by an unusually wide range of the populace.

Thus Professor Shragga Irmay, of the Israel Institute of Technology, was able to explain how Israel learned from the experiences of more advanced countries. He said: "In highly developed countries, such as the United States, the United

Kingdom, and Germany, technology and educational methods grew up gradually for centuries, in parallel with the development of their industry and structure. This was often achieved by the wasteful method of trial and error and by destructive wars. Europe has not yet overcome the problem of cramped and tortuous cities and of economically unsound distribution of population centres. Engineering education in Israel has not developed exactly along such lines. The Jewish immigration, mostly from eastern and central Europe, brought in a number of engineers who founded in 1921 the Association of Engineers and Architects of Palestine."

He continued: "The rapid settlement of Jews on the land called for drainage of malarial swamps, installation of modern water supply and sewerage systems, construction of roads and buildings. In 1924 the Technion was founded. It is now the Israel Institute of Technology, overlooking the populous and industrial Haifa Bay area. Technion then turned out civil, mechanical, electrical, and chemical engineers, who, while professionally competent, had a narrow scientific base. In 1948 the establishment of the State of Israel and the explosive fivefold increase of population by mass immigration within 20 years, stimulated the development of Technion and the addition of further engineering departments—agricultural, aeronautical, mineral, bio-medical, industrial, and management engineering."

Last year Israel's universities and higher institutes had 4500 academic staff members and 28,500 students. Of these, 17 per cent studied engineering and agriculture, 17 per cent mathematics and natural sciences, 4 per cent medicine, 23 per cent social sciences, 32 per cent humanities, and 7 per cent law. Israeli researchers are so skilful and diligent that last year they contributed one in 200 of all scientific papers published in the world, although Israel's population is only one in 1400 of the world's peoples.

Although sciences and medicine are taught at other universities, such as the Hebrew University and those at Tel Aviv and Bar-Ilan and the Weizmann Institute, and new universities have been established at Haifa and Beersheba, more will be needed soon.

But Professor Irmay thinks that in addition to local training, "bright young people should be sent abroad for training and education to serve later as technological leaders to their countrymen". He added: "It is important to choose emotionally stable candidates for this kind of scheme, and to send them to those countries and places where they will be received in a friendly way, without desiring to remain there permanently. For each engineer at least three technicians should be educated. If this is done gradually with curricula and methods adapted to local needs, without excessive red tape, the selection of candidates and personnel made on ability and not on extraneous criteria, then success is assured. In such a way, we hope to avoid the 'brain drain' experienced by other countries."

And at the bottom end of the educational ladder, Mr. D. Gilboa, an official of Israel's Ministry of Defence, informed the conference that the Israeli Army had its own teams of educators who were helping to cope with illiteracy among recent immigrants to the country.

Professor Ben Roberts wanted to know how the distribution of students between the various categories in higher education had been decided. Was it related to some manpower forecast for the future? Was it by chance? Was it the right distribution?

Later, Professor Irmay confessed, "The most surprising thing is that, as far as I know, there was no government policy concerning the distribution of academic manpower, students, and so on. This was done, I would say, haphazardly, but somehow we managed just to supply the needs of the country. In a certain sense the leaders of the universities had that foresight without having all the necessary data. It is difficult to make forecasts and predictions these days, even in the field of economics; how much more difficult in the field of education, where the results are felt 10, usually 20, years later."

Israel, however, still faces the problem of integration. Professor Irmay explained: "We have a lot of illiteracy. We have to teach people what is electricity, what is public transport, what is soap, what is toothpaste, what is eating normal bread, and all kinds of things they never saw. Mainly the illiteracy is among women who came from the Islamic world. For example, all Jews who came from the Yemen were 100 per cent literate, but their women were 100 per cent illiterate. This is what we call the melting pot, and it cannot be solved in one generation—not even in two. We have the problem of integration, I hope not as acute as in the United States."

There was another point of topical interest to the Western world. Professor Irmay commented: "If we do not have student unrest, this is due to the fact that we all have a common goal of survival, which is very important."

Mr. Yosef Ami forecast that more research would be undertaken in Israel on formal and informal education to meet new situations. The planning of education was the most long-range activity of human management. He explained: "The age of automation will be the age of permanent and continuous learning for all of us."

He continued: "A permanent and continuous learning seems to be a more exact definition of our future lives. Education and the training of manpower is a matter of national planning. Taxes and levies on income and products to create the funds needed for adult education and training are probably the best solution of the problem. This makes the education of adults, training, and re-training a social "must". But it will not help us if, in our managerial jargon, we say this is a problem of government and not of local management. The problems are there at plant level, and there is no other way to solve them than to become more human-factor-minded, and to see a central activity of ours as educators of the people we are in charge of."

Mr. Ami, who is vice-president of Technion, suggested that man would accept the task of learning through life if he realized that only in this way would he improve his income, his status and his security.

A fascinating glimpse into the growing zeal for education in Israel's *kibbutzim* was contributed by Mr. Dan Ram, a former *kibbutz* worker who is now an official of the Ministry of Labour. He explained that the *kibbutz*, which was formerly a wholly agricultural society, has now developed industrial interests, and in some *kibbutzim* industry has grown to 40–50 per cent of the total turnover of their economic activities.

Mr. Ram continued: "This physical change has been followed by a change in mentality. From an ardent opposition to matriculation examinations as an integral part of high school learning, which takes 12 years, opposition to sending sons to an academic institution, and even an attitude of ridiculing the academic certificate—if you wanted to abuse someone you called him an 'intelligent'—we now find people of the *kibbutzim* learning in the University, the Weizmann Institute, and the Technion. A vast system of variety of courses is carried out by the *kibbutz* movement itself. There is even talk today about a *kibbutz* university. It is anticipated that about the year 1980, 20 per cent or even more, of the adult population in the *kibbutz* will be graduates of academic institutes. This is an example of quite a revolutionary development that is taking place without any serious crisis—except lengthy, tiresome discussions carried on to the small hours of the morning!"

He conceded, however, that in educational facilities for the young there was a conservative attitude and opposition to change. Mr. Ram insisted: "It is the duty of education to enable the child to acquire the ability to learn by himself, to acquire self-assurance in his capability to do it, and to make it his habit. There is no doubt that children who are used to think and work independently at an early age will be able to cope better with a changing environment, and ready to take responsibility in unpredictable situations and under stress conditions."

Mr. Ben-Ami Friedrich, an industrialist, observed: "The question of learning, of permanent learning, is also a question of societies. You have societies where this is possible. For example, it is ingrained in the Jewish society for thousands of years to learn and study from childhood until a high age. But I wonder if other societies which are now emerging as developing countries will also be able to get into these habits? Another factor which disturbs this question of continuing study is the outlook towards free, or leisure, time."

Dr. J. N. Peri, of Tel Aviv University, thought that Dr. Clark must have been desperate to have suggested giving industry *carte blanche* to take over complete responsibility for education. The problem, he claimed, was not so much entrusting industry with education as permeating the educational system and educators with "some of the know-how, the knowledge, the tools, the zest, and the methods which made industry and commerce in the United States, and in other countries, so prosperous and so efficient. I believe that is where we should put most of our effort."

But Dr. Peri probably delivered the most appropriate postscript to the whole problem when he concluded:

"I am not so pessimistic as far as education is concerned, either in the United States or in Britain, where I had the privilege of spending a few years—or, naturally, in Israel."

British participation in the discussions of educational problems was not as prominent as that from either the United States or Israel. The greater concentration of the British team was reserved for some of the more technical aspects of the programme.

There was a strong academic representation from Britain, and there were enough references to education to reflect an acute awareness of its importance.

Professor Crossley, for example, wound up his treatise on the sharing of capital gains by declaring: "Few governments have yet begun to comprehend the full scope of their powers and responsibilities with regard to what is probably the most important single factor influencing the sharing of income, which is the distribution among the population of human capital in the form of education and training."

And Mr. John Hargreaves, commenting during the discussion on the effect of technology on the dispossessed, referred to the fear so many people had of what was termed a "dictatorship of technocrats". He recognized that the language and the particular expertise of people engaged in technology seemed to place them in a world to which ordinary people do not aspire. But, he added, "the only guarantee of a democracy of technocrats is that ordinary people have a basic education in technology".

Professor Kaldor, too, concluded his paper on underdeveloped countries with the declaration: "One cannot emphasize too strongly that the efficiency and speed with which modern technology can be introduced in underdeveloped countries is very much a matter of the quality and character of the system of education. In many underdeveloped countries, too much of the resources are spent in providing high level education of a theoretical kind, and too little in creating a cadre of qualified managers, engineers and technicians."

And Sir Denis Barnes reported that the British Government is expanding its programme for providing training and retraining centres "for all levels and all ages".

Some reservations about the panacea of education, however, were expressed by Mr. Patrick Fisher, of the TUC, when he said: "I think, of course, that it is extremely important that educationists and those concerned with vocational training do their best to ensure that workpeople, and particularly new entrants into industry as well as those already in it, can adapt to changing job requirements. But I think that, perhaps, it is a little too presumptuous to imagine that this vocational training and these new ideas about education, continuing the life-long educational process, is an absolute necessity. We tend to overlook the very great adaptability of the human being."

He continued: "I think the surprising thing to many people is the way in which coalminers, for example, have adapted themselves very quickly to quite advanced forms of technological employment, with very little vocational training. What seems to be important is that there should be a background and a basis of a good educational tradition, on which people can make the adaptation. I am not altogether convinced that vocational training has to be as detailed as some people tend to think."

At the end of the conference Mr. Fisher recalled that it had been estimated that it might take up to 25 years or so to overhaul educational systems. He commented: "If that is so, it appears to be a good deal slower than the technological changes which are now taking place."

Mrs. Joan Blakeman (Miss Joan Woodward), of Imperial College, and a consultant to a number of industrial companies, reported that few problems had been met in industries known to her in retraining workers inside advanced automated factories.

And Mr. Edwin Nixon, who is the head of a computer manufacturing firm, confirmed that there was little difficulty in retraining people "if you go about it in the right way".

He continued: "In training for the future we should certainly intend to train our people in whatever new technologies develop. We may be unusual as an industry, but in the United Kingdom, with a work-force of 10,000 we have 250 trained teachers. We have talked about education. I do believe that there is a need for changing the basis of education, so that people coming into industry can have a basic understanding, or a basic logic, if you like, on which you can build a particular expertise."

Mr. Nixon concluded: "I am certain that the new mathematics programme in the United Kingdom is leading to the development of people who will be able to accept new technologies."

There appeared therefore, to be an uncritical attitude by the British participants to their own educational system. There was not even a murmur about the lively current controversy between the government, local authorities, and teachers; not even a whisper about the continuing goings-on among British university students.

Such reticence underscored the slightly envious comment of Mrs. Mary Conway Kohler a day or two later, when she observed: "The British are so reserved—they don't expose their soiled linen quite like we do."

Chapter 4

Tomorrow's Men of Status

ONE of the most intriguing glimpses of a possible future world was provided, perhaps somewhat surprisingly, by the traditionally realistic Sir Denis Barnes, the respected British civil servant. In effect he permitted us to share in his "thinking aloud" on the prospect of a society where half the populace would accept with contentment the notion of endless leisure.

To be fair to him, Sir Denis was not making prophecies as much as inviting those who read his paper to cast their minds to far horizons. And it seemed that he became a trifle startled by his imagination, for he suddenly concluded: "I hope I will not be misunderstood by ending on this futuristic note." And he added: "I do not wish to underrate in any way the significance of the problems of the present and immediate future. All I suggest is that we look at them in perspective."

But this was his engaging flight of fancy. I record it fully for two reasons. Its vision; and its contrast to the realities of earlier tomorrows.

Sir Denis wrote: "Looking well into the future is it quite unreasonable to think that we may be in a situation somewhat as follows? A much smaller proportion of the population of working age will actually be working—it could conceivably be lower than 50 per cent. The number of hours worked per week will be much less. Nearly all manual jobs as we know them today will have disappeared. Jobs calling for physical effort will be done by machines; all routine clerical work by computers. Employees will be engaged either on very high level managerial and technical work, or as machine minders. People will start working much later in life and retire much earlier. Working lives could conceivably average only 15–20 years. The average income of all adult persons in the community will be much higher in real terms than it is now. Today's concept of 'unemployment' will have gone completely.

"While in the relatively short term the process of finding people jobs and making the best use of our manpower may become more complicated, we could in time face the entirely different problem of making routine work (and leisure) more bearable. For various reasons trade unions and employers and their associations may fade away. Comparatively few people will be directly involved in the community's production and commercial system. The advance of information technology will give more objectivity to all social and economic decisions—the relative subjectivity of these today is one of the essential reasons for the existence of the employer–trade union system. Advances in technology and management methods, and the changed nature of the work force will mean changes in the organization and management of undertakings and work. There will be a closer relationship between the State and the giant industrial and commercial undertakings. Perhaps few of these changes will take place. Even if they do it is impossible to say in what time scale. But I think some are possible enough to be given serious thought."

And that was the vision of Sir Denis Barnes. Understandably, conference participants were too intensely preoccupied with more immediate preoccupations. Yet there were glimpses during the week of the impact made by this idea of an ultimate Utopia.

For instance, in the very last session of the conference, Dr. J. Wilhelm Kluver, of the United States, wondered about the problem of educating South America and India through satellites providing a television system, which could be a fact in 20 years. He asked: "What are you going to teach people to do? Maybe it is possible to say that any kind of activity is work. Is it possible to incorporate activity into economics—say ski-ing or scuba diving, and so on?"

Mr. John Hargreaves, of Great Britain, replied: "Certainly one of the effects of the type of society that is evolving now is that the last century's somewhat Puritan relationship between toil and income is going to disappear, and therefore one has to think in wider terms of activity than work. Certainly, too, we have to ponder education, not only for a job, which is the present trend, but education consciously for what one may call non-work. One has to find the moral equivalent for work."

But, alas, long before that time the impact of technology threatens to create many conflicting and harassing problems demanding foresight and energy, patience and adaptability, toleration and efficiency. Clearly emerging are two distinct aspects; the role of those who will work in the modern, highly technological factories of the more immediate future, and the plight of those for whom no place can be found. While there was a hint of despair at the magnitude of the tasks ahead, particularly in the United States, there was also a courageous determination to devise immense schemes for the absorption of the industrially unwanted into work of communal usefulness and personal dignity.

Mr. P. L. Siemiller, leader of the International Association of Machinists and Aerospace Workers, warned that today technology is challenging many of the basic ethical concepts that have developed out of the Judaic—Christian tradition. He added: "Such innovations as the bomb, the computer, the satellite, the pill, and the organ transplant are revolutionizing ethics as well as technology."

He recalled that there are those who view present technological change as a forerunner of future social catastrophe, including Robert Theobald, who foresees a future in which only 5 or 10 per cent of the adult population will be usefully employed while the other 90 or 95 per cent exist idly and aimlessly on some kind of dole.

Mr. Siemiller added: "This is a gloomy forecast. If technology should displace people to such an extreme the results for the human spirit will be catastrophic. I do not believe that technological advance warrants such pessimism. I believe that collective bargaining can help to adjust the work load to the available work force. The rewards of increased productivity can be distributed in many ways—not only in higher standards of living through higher wages but in more time for living through shorter hours. There is no doubt in my mind that automation will make it possible—if not imperative—for workers to work fewer hours in a week and in the course of a lifetime."

He continued: "There is nothing sacrosanct about the 40 hour week, the 2–4 week vacation or the 62–65 year retirement age. As time goes on unions in the United States and elsewhere will negotiate more urgently for shorter work weeks, more 3-day weekends, longer vacations, and earlier retirement voluntarily.

"These are the goals of collective bargaining that will help to reduce much of the tension that technological change might otherwise create between labour and management. Finally, though I do not subscribe to the optimistic doctrine that computers will create more jobs than they destroy, I agree that they can release people for jobs that are very much needed and very much neglected—in our society. I am thinking of services for the old, the sick, the young, the physically handicapped, the mentally retarded, the emotionally ill, the psychologically disturbed, and the socially alienated. In the United States, despite the fantastic wealth that has been created by America's industrial capacity, these large areas of social service and national concern are tragically understaffed, undertrained, underpaid, and underdeveloped. Hundreds of thousands, possibly millions of job opportunities can be created in these areas of social service alone. And, of course, there are many other jobs to be done—both in America and in the world—to create a richer and more rewarding environment of life for all men. The question is not whether there will be enough jobs. The question is whether—in periods of technological transition—people can be shifted from old to new forms of useful employment with the least possible pain—both to themselves and to society."

Mr. Siemiller concluded: "The periods of transition will not be easy. They never are. But with the help of collective bargaining

—which was developed to reduce conflicts generated by the Industrial Revolution—labour and management can work together to reduce tensions created by the computer revolution."

There were far more passionate undertones in the contribution of Mr. Whitney M. Young, the Negro leader and executive director of the National Urban League.

He declared: "Our age is not unique in either the swiftness of technological change nor in its impact on our society. It is unique, however, in its ability to control the effects of such change to ensure that it is used to benefit a society and to cushion its effects on those least able to adapt to change."

He continued: "The effects of industrialization on nineteenth-century Europe and America, and on some countries in our own time, have been disastrous for workers, lower-status groups, and nations unable to defend themselves from the plunder of their neighbours. It wasn't only the British weaver who suffered the effects of technological change in the nineteenth-century textile industry; it was also the black African who was sold into American slavery, the profits of which were invested in European industrial expansion. The developed nations that are rapidly altering the way they produce goods and services must be aware that they have the capacity, and must develop the will, to prevent their burgeoning technology from becoming a Frankenstein's monster. This is especially important when we realize that the inroads of technology will increase at a far faster pace in the future. It is estimated that by 1975 the United States will be spending more than £14,000 million on research and development—in 1940 it spent a mere £95 million."

Mr. Young wrote: "As in any period of accelerated change, there will be dangers as well as new opportunities for our society. Technology holds the promise of freeing man from the tedium of repetitious, alienating work, and of freeing his energies and his spirit for the creative and satisfying tasks that affirm his humanity and his labour. On the other hand, it can destroy the fabric of democratic society by concentrating power in an educated élite who, possessing the computerized tools of the new technology, will insecurely rule over a mass of dispossessed citizens, excluded from the fruits of the new era. We possess the tools to make technology work *for* us, and if it turns into a nightmare, it will not be the fault of technology but of the greed and stupidity of men."

Much of the problem, he suggests, centres about the capability of governments to use technology to create new jobs for expanding work forces, and especially to deal a death blow to poverty by guaranteeing jobs or decent incomes to all.

Mr. Young argued that in the United States this challenge takes the form of providing social and economic equality for Black citizens and other ethnic minorities who suffer disproportionate poverty.

The impact of technology is especially disastrous for these groups, he asserted. Lacking the skills and education required for more sophisticated jobs, they are disproportionately concentrated in the very unskilled jobs most likely to disappear through the introduction of new machines.

Mr. Young is obviously not discouraged by the projections of the Harvard economist, Dr. Otto Eckstein, that by 1985 Negroes in the United States will hold a smaller percentage of jobs in professional, skilled, clerical, or sales fields—the very areas in which employment can be expected to grow.

But Mr. Young concedes that the impact of technology on the dispossessed is not solely a racial problem. The White poor, he declares, far outnumber the Black poor, and far more White workers are unemployed, underemployed, or on welfare than Blacks.

He added: "Whether we look at entire regions, such as Appalachia, or large cities like New York, where one out of every eight people is on welfare, White people form the bulk of the dispossessed. If I dwell on the problems of the Black poor it is because their concentration in the metropolitan economic centres of America makes their poverty, exacerbated as it is by persistent discrimination, the major problem facing the American economy."

Mr. Young believes, nevertheless, that Negroes are concentrated in jobs that are most prone to technological displacement or unemployment, and within those occupations suffer disproportionate poverty. The outlook for future employment opportunities seems restricted to occupations demanding maximum education and training. He estimates that the largest increase in job opportunities—45.2 per cent—will be in professional and technical jobs, while the greatest decrease in employment will be among farm labourers, further intensifying the migration to the big cities from the rural south.

What then is the basis of Mr. Young's hopes for the future? He cites the projected changes in employment in major industry divisions as reported in the Manpower Report of the President, 1967. These anticipated by 1975 an increase of 12,421,000 in the service-producing industries.

And he comments: "If ways can be found to fill these projected job opportunities in service-producing industries with workers displaced from goods-producing industries where productivity comes increasingly from machines rather than human labour, we can assure that the new technology will work *for* men and will not increase the ranks of the dispossessed. About two-thirds of the service sector's increase in jobs will take place in government (especially in non-federal employment) and such callings as health services."

Mr. Young asserted: "It has become a truism that as private wealth and spending expand, badly needed public services lag. The sheer backlog of unmet needs in public housing, education, and health is staggering, and the potential needs of a larger, still more urbanized population in the future make it apparent that public employment is the answer to the manpower needs of the nation."

Mr. Young was a member of the Commission on Technology, Automation, and Economic Progress. Its report in 1966 identified about 5,300,000 new jobs that could be created in medical institutions and health services, educational institutions, national beautification, welfare, and home care, public protection, and urban renewal and sanitation.

He now insists: "There is a precedent for such massive hiring in public employment. During the depression, millions of workers were hired for work relief and public works projects. Thirty years ago, 6.6 per cent of the labour force worked in such programmes and the cost was 2.8 per cent of the gross national product. By comparison, a rational programme of public service employment for the 1970's would be proportionately less costly, while fulfilling unmet social needs and bringing the dispossessed into roles as producers in an expanding economy. The short memories of now-affluent Americans have transformed the job programmes of the thirties into a myth of millions of men raking leaves and shovelling air. But few Americans have not driven on roads built in those programmes or used public buildings and facilities built by the WPA. More than 650,000 miles of streets and highways were built; 35,000 public buildings went up, 77,000 bridges and viaducts were erected, 3000 playgrounds and 3000 athletic fields along with 2800 utility plants were built. The cost—10,000 million dollars—was cheap for the increase in public facilities and the relief of severe unemployment.

"Today's needs are no less pressing. Along with the existence of an urban underclass suffering the indignities of poverty, unemployment, and underemployment, there are overwhelming unmet needs for expanded social services. Certainly the mass education so vital to a technologically advanced society has sharply increased the need for workers in schools, while the increased productivity that is a feature of the new technology has resulted in more leisure time, which means increased use of public facilities such as parks, rivers, and libraries. A large part of public needs can result in job opportunities for labourers and unskilled workers. The bulk of future public employment will come in such jobs as housing and pollution inspection, hospital workers, fire and police services, recreation and parks, education, and similar areas where overburdened professionals need assistance from sub-professional workers who can carry out essential jobs for which technical knowledge is not required."

Mr. Young concluded this theme by saying: "Public service employment along these lines is a necessity for a nation that needs to upgrade inferior public services and meet the needs of the future, and it is a necessity for a nation unwilling to face the consequences of the existence of a mass underclass, in effect a tinderbox of explosive urban anger. Expanded private employment is necessary, and incentives should be given to industry to hire and train the unemployed and to restructure jobs to allow for greater worker satisfaction and mobility, but the combination of the need for public employment will be the route America takes."

But Mr. Young also introduced another theme which has significant implications, and inspired comments from many subsequent speakers at the conference.

He explained: "There is one other aspect of the problem of technology and the dispossessed. It may be seen more as a political one far removed from the mundane discussions of manpower required and job categories, but it could be the major aspect in the coming decade for a society caught up in rapid social change. The increasing impact of technology on the economy means that the power relationships within society are altered. The existence of a new class—called "technocrats" by some observers—who hold the educational requirements to control the new technology poses new questions of how decisions affecting masses of people are to be made."

He continued: "Decisions once made by political office-holders accountable to the citizenry are increasingly being made by technical experts accountable to no one. Whether to build a road here or a hospital there is too often decided by experts who arrive at their conclusions on the basis of data fed into computers, without the necessary regard for the intuitive feelings or demonstrable interests of the people affected by such decisions. It is entirely possible that when land costs, traffic flow, and a multitude of other data are processed, the most logical route for a new road will be through a slum area. But unless the 'experts' take into account the social costs, not so easily quantifiable, their decision may precipitate the destruction of a stable neighbourhood; the homes and businesses of countless citizens who have had no power to prevent the unwanted violation of their lives."

Mr. Young pointed out: "Technology makes possible the existence of an élite in effective control of the decision-making apparatus of a society awed and bewildered by their technical expertise. It may be argued that most of their decisions are really made by politicians or public officials, and that the 'technocrat' merely advises. But few people can stand up to 'scientific' evidence or 'expert opinion', and when the experts are divorced from the realities of the everyday life of the dispossessed, the way is open for tragic abuses of both knowledge and power.

"Clemenceau said that war was too important to be left to the generals, and that observation holds true in the social arena today. Technology fosters centralization of power at the very moment in our history when the dispossessed are demanding a role in the decision-making process for themselves."

He declared: "If technology is to work *for* man society must avert the danger of creating an undemocratic élite whose devotion to 'objectivity' and scientific method serves to lock the dispossessed out of effective control over decisions affecting their lives. Failure may mean that the ground will be laid for social dislocations that could destroy an urban society."

Some of the most perceptive observations on the effect of technology on human society came from the Israeli industrialist, Mr. Ben-Ami Friedrich. Though the general tenor of his paper on "Changing Skill Requirements, Job Training, Motivation, and Mobility" was detached and philosophical—"changes take place in every era but they have not yet caused a revolution"—he recognized that automation is liable to give rise to weighty interferences in the structure of the economy and in the traditional "prestige" strata in the professions and in industry. "This transit", he added, "will last until young people have been assimilated in the new vocations and older people leave the economic cycle."

And in an arresting, though somewhat forbidding, phrase, Mr. Friedrich referred to "the generation of the wilderness".

They are people in the age group between 35 and 55, whose original occupation will be down-graded by automation. Presumably he regards most of this group as a bit too old to learn new skills and a bit too young to retire.

But here is the submission of Mr. Ben-Ami Friedrich: "In all periods of human culture there have occurred, frequent clashes arising from lags between the achievements of the human mind and human capacity. In my opinion there is no escaping from the fact that with the advance of automation only a certain portion of the manpower releases will be able, after systematic re-education, to integrate into the industrial process.

"There are many doubts as to whether the 'capabilities' existing on the lower vocational levels, which are generally a consequence of a lack of intelligence, will indeed be sufficient to overcome, even by suitable training methods, deeply ingrained habits of people of a low educational level and above a certain age, and this in addition to the tremendous psychological barriers that cancel out almost all efforts being made on their behalf. It is most difficult for grown-up people to admit that they have in fact for their entire lives engaged in the wrong occupation and to recognize their social inferiority. It was the feeling of being necessary that really gave these employees the confidence that they, too, were indeed worthwhile members of the society in which they lived. This situation applies even to many educated people in administrative occupations, and what may be achieved with these, on the one hand, by virtue of personal willingness for advancement, will be lost, on the other hand, by the fear of descending in the social scale.

"There is in general the question of whether in the framework of administrative practices it is possible to solve problems such as these in their entirety. It is perhaps necessary to conduct practical experiments in regional frameworks or even in a limited economic environment in order to find a temporary balance between various 'skill levels' and the needs of individual plants. Within this framework the manufacturers would have to bear the burden of the costs of the manpower that could be fitted in by raising prices in the wake of increased production costs, or the government by raising tax levels to support subsidized workers, following their ejection from the work cycle. In any case the ultimate consumer would of course pay the bill.

"Let us assume that within a short time we shall arrive at a balance that will be tolerable to the establishment in most respects, and from the financial standpoint too. It is supposed we shall arrive at this balance by natural vocational selection and departure from the work cycle through death or retirement, on the one hand, and through productive employment of part of the manpower on the other. I should then like to dwell upon one point which has perhaps greater weight as regards the level of employment than the question of changing the level of skills required. We should view with great concern the sharp social polarization which can be expected in the wake of the changes that automation will bring.

"Talented persons will be drawn by the higher requirements to the professions, which will simply be closed to the less talented. They will undoubtedly form a new 'caste'—the productive caste, productive not only from the simple standpoint of production of commodities, but productive from the standpoint of managerial, decision making, executive and control systems, whereas the larger number of less gifted, some of whom today are undoubtedly integrated in the conventional or half-automated management processes, will within a few years be absorbed by the services. These will form another 'caste'—the caste of servants.

"The conversion of large groups to the status of 'servants', even though they may be given some of the halo and the glamour of economic branches such as entertainment, fashion, tourism, transport, etc., will change the accepted order of social strata. Because of the different educational requirements for the two sectors—production and services—a gap will be created which will undoubtedly have more and more of an effect on the class trends. This is something we are not accustomed to. It is something that we have not known except in the times of aristocratic regimes. It does not matter if the material differences are not large. What will be decisive in the matters of caste will be professional ability. Those who have such an ability will be the rulers in the establishment. It may also be that among those of a high level will be included, too, those parts of the population which until now have comprised the so-called humanistic class—such as teachers, doctors, etc., who will be more and more affected in their work by the mechanization that is also penetrating these professions.

"Only with special efforts in the area of education in general, and in the area of vocational education in particular will it perhaps be possible to achieve moderation and balance—though shaky—between polarized vocational groups. Is it possible then to assume that by instilling suitable motivations, or by advancing the willingness for changing planes of skill, some relief may be hoped for? It is undoubtedly possible to achieve something. But I might ask with some scepticism whether human motivations also rise to the same level of progress as does the technological challenge, although the aspiration for material improvement might still be satisfied if we assume the egalitarian process that has marked the post-war period. But it will come to an end by the polarization of activities and will probably give way to both a social and a material differentiation. On the other hand, the desire for social status will lead to frustration which could reach such proportions that even more sophisticated programmes for increasing motivation will not prevent a negative reaction.

"In those few cases where an attempt has been made to motivate a plant to 'automation consciousness' by incentives, there has been perhaps an initial success. It is too soon as yet to reach conclusions about such measures. They have yet to prove themselves effective in a broader framework. They are of value within the individual plant, but then again motivation is of value in every form of management, and it therefore has no special validity in a specific field such as automation. In summing up it may be said that the employment of motivation will not be very effective owing to the polarization of new occupational classes. The preferential professional branches that will 'sink' because of automation to a lower skill level almost can not be moderated to accept the new status, neither material nor social. Even if the new vocational strata are given a wage level that offers them rising standards of living, I do not for a moment see the way to utilize this as a means of introducing automation, and thereby to relax and limit the areas of tension that might be expected. We should not forget that the conservative forces within us are stronger than promises of good times that are not so certain. The fear of the unknown, the hesitation to face new trials and new challenges, are more weighty by many times than the 'carrot' that is termed motivation. I would never deny its value and far be it from me not to use it, but as a solution for the overall problem it will be of small and not of decisive value."

Mr. Friedrich, challenging and provocative, urged leaders of the economy and of workers' organizations to view these aspects in a broader framework and in proper perspective, and to free themselves to a great extent from prejudices and dare new approaches with new ideas, "without a burden of 'sacred cows'", which were, he submitted, "only the idealogical crystallization of momentary ideas", which he hoped would become obsolete even more quickly than modern technical systems. He acknowledged, though, that "the inertia of old ideas in opposing innovations is formidable."

Mr. Y. Meshel, one of the leaders of the Israel General Federation of Labour (Histadrut) confessed that the trade unions have been looking on the first steps of modern technology "with one smiling eye and with tears in the other". On the one hand, brilliant perspectives opened towards improving working techniques and bringing about a general rise in the standard of living of the workers; on the other hand, a great concern had spread on the grounds that advancing technology would cause hopeless mass unemployment.

In his view, the sharpest social problem in advancing technology is the possibility of confrontation between the veteran workers and the young ones.

Mr. Meshel continued: "It would be a very bitter dilemma for the trade union if it should have to defend the young man against his older colleague who has become deeply rooted in his working place and has acquired a sort of ownership over his job, when all of a sudden the young man, who has just finished his studies comes in and wants to push him aside. Nor would it be less difficult for the trade union to back the veteran worker against the young man, who has acquired up-to-date knowledge and is yearning to make use of it with youthful energy, yet the place he aspires to attain on the strength of his professional training and juvenile dexterity is occupied by someone who is older in years but poorer in professional knowledge and skill."

This situation could be ameliorated, Mr. Meshel hoped, by a combination of vocational retraining of older workers, acceleration of vocational mobility, and early consultations between employers and unions.

But it was the "élite" theme propounded by Mr. Friedrich which stimulated many comments.

Mr. Yosef Ami, vice-president of the Technion, Israel's famed educational institution, stressed that education was the surest protection against inevitable changes wrought by technology. He declared: "Technological progress and automation mean facing permanent change in our working place, in our communities and in our nationhood. The age of automation will be the age of permanent and continuous learning for all of us. You may call it, if you prefer, training, study refresher courses, brush-ups, and so on. But in the end it is permanent learning and permanent education. A permanent and continuous learning seems to be a more exact definition of our future lives. Education and training of manpower is a matter of national planning. The whole social system ought to become active and interested in the general effort of manpower development. Education and training cost money and who should pay for that? It is today recognized as a social overhead, and the public should pay for it. Taxes and levies on income and products in order to create the funds needed for adult education and training are probably the best solution to this problem."

It was generally accepted that at a certain age learning stops.

Mr. Ami retorted: "I have seen my grandfather sitting at the age of 90 studying the Talmud. If he could do it, I don't see why at the age of automation we can't study the 'Talmud of Automation' at high ages, too. Man will do it if he will be conscious of the fact that only through permanent education and training will he be able to improve his income, status and security. This is the problem—to create that direct psychological link between permanent learning and progress in our own individual affairs as part of the group in which we act."

Mr. Dan Ram, who spent many years in developing the industrial sector of Israel's *kibbutz* movement and is now one of the assistants in charge of vocational training at Israel's Ministry of Labour, warned: "We should be very careful in predicting disastrous developments from technology. One of the predictions which I consider a very dangerous one is that of the formation of two distinct classes—the technological and scientific aristocracy and the caste of servants. Prophesying such a future is kindling and inviting ardent and passionate opposition to change and progress, and turning the natural reluctance of the human being towards change to a new sense or instrument of restraint against new developments. The prophecy of social polarization and the fact that it takes place in the United States is not a divine law, and it need not happen everywhere. Although there is mutual influence between social polarization and vocational training, they are not synonymous, that is a reflection of the virtues of societies and their organizations."

Mr. Ram continued: "As technology and automation proceed, we are becoming more dependent on each other, but while dependence causes frustration, mutual dependence fosters responsibility and diminishes social barriers, and I think acts against social polarization. It is true I think today, even in such an organization as the hierarchical establishment of the army, that it cannot keep high morale, efficiency and devotion when there are snobbish or artificial barriers between the ranks. The more the army is democratic, the more it is efficient. I can think today of something that epitomizes this development. I know in fact that a boy grown up on a farm can find a common interest with a scientist from the Weizmann Institute, crawling in caves and looking for flowers or studying the life of bats, and they just come together on the same rank and level. I know that this is a fact of life because I refer to my son and a professor from the university. This is not a single phenomenon. The versatility of activity in human society is so great, and there is now more opportunity for the individual to find an outlet for his needs and counterbalance shortcomings in other fields. The duty of society is to equip the individual through proper systems of education and methods of vocational training so that he can be more flexible and free in his choice."

Another Israeli participant, Mr. A. Shemer, also demurred at the idea of an élite and "caste" of servants. He declared: "As for polarization, I too found it very difficult to accept this division into the creators and the servants, even if it was written in inverted commas. If we bring it in this way, it is very difficult to absorb and accept, because those who have been raised in a free world and a world of free ideas that placed man in the centre of things cannot accept such a division. We reject it."

Mr. Shemer continued: "Let us think for a moment why precisely in this age of automation polarization should grow. Today, too, there are directors, there are rank-and-file workers. It is true that we have this kind of distribution, this division. We have heard in the past, and we have continued hearing today, that the division is between policy makers and policy executors. But even if we do accept such a division, it is a long chain, there are many links and many colours. I think that in another period, the one towards which we are advancing, with the change in various occupations, it will be impossible that this world of automation will be divided in two only."

He concluded: "If Mr. Friedrich's assumption is correct, I see this as a source of danger for our existence. The education which we advocated said, 'every man creates in his sphere, even if this is a narrow sphere'. It is creativity that matters, and it cannot be the sphere of a sect or a unit. It is a gift of nature, and we as a democratic and free society must preserve it at all costs."

Mr. Friedrich replied that he would not like to see society based on two castes. But, he added, "I see the danger in introducing automation without thinking of these problems. It can be dangerous to prophesy on this, and I did not try to prophesy, but I think that we must talk about it, we must see it very clearly and maybe very cruelly. We will face these things if we will not try to adapt ourselves to these new times."

He continued: "The question of building an élite and upgrading—I think this is quite a natural development. And when I remember when we were boys and when we thought the highest grade of professional standing was to be the driver of a locomotive, then it changed to be a pilot, and maybe tomorrow it will be the researcher or the scientific adviser to the government. But we always think in upgrading ourselves. And would you not find today what I have mentioned? I don't believe that we will find in the young people within the next 10 years, when they finish school, that they will want to be just simple accountants, writing figures in a ledger book. I don't see them. They may go into programming of computers, which is already a much higher level of professional work. You can call them servants, you can call them creators, I wanted just to show that we have two different lines of young people advancing towards their professional lives. It is already difficult to find young people coming and wanting to do physical or simple work. They all want to be at least in the maintenance shop—it's a kind of upgrading. I didn't mean to say that servants are serving the others. They are serving the community. And as such they are humanly, socially or morally on the same level."

Maybe the vision of Sir Denis Barnes which implied the blessings of a time when machines can be subjugated to the service of mankind is too remote to act as a spur and an inspiration to hasten the conquest of intervening phases.

It could be that Mr. Whitney Young is over fearsome of the powers of the technocrats; and that Mr. Friedrich is over-simplifying the emergence of a new two-caste system in technological society; and that other speakers were too complacent in their confidence that trade unions, employers and their organizations, and governments will be able to "iron out" the problems which are bound to come in the wake of massive technology. But all the exponents felt justified in drawing attention to their hopes and fears.

There appeared to be a concensus on two issues; the first is the importance of education, even if there are differences in the form it should take, and the second is the advantage of being youthful and receptive, younger than "the generation of the wilderness", whose penalty in many cases will be upheaval from the routines they have known and whose salvation will be adaptability.

Chapter 5

The Future Role of the Unions

IF ONE accepts that the pace of technology is accelerating and men have to adapt themselves more extensively than ever before to new conditions, then, surely, there can be few institutions in more urgent need of reappraisal than the trade unions. And, for that matter, associations representing employers, though these vary so much even in individual countries that it is difficult to discern for them any coherent pattern.

It has become an accepted habit to regard the trade unions in advanced countries as either an inevitable feature of society or a necessary evil. Few of their adherents would regard them as perfect, and they have many critics reluctant to concede them any virtues.

Yet their importance is undeniable; but being the creations of man they can, and do, reflect the whole range of human attributes. The variation in the standards of their performance and in their conception of their ultimate objectives is immense; their performances can sometimes be wildly different from their intentions; and their ideals have been known to be undermined by expediency and selfishness. Even so, their achievements have been numerous and impressive; their strength has been both an instrument for the determined and a protection for the oppressed, and it is almost unthinkable to visualize an industrial society without them.

From time to time a trade union leader may give, unintentionally or by design in special circumstances, an impression of arrogance. It happens rarely—but rare, too, is the salutary reminder which can be accepted with grace that unions must guard against an assumption of inviolable indispensability.

The diminutive stature of Dr. Rivka Bar-Yosef is in inverse ratio to her intellectual dynamism and her authoritative candour. She is the chairman of the Department of Sociology of the Hebrew University in Jerusalem, and a doctor of philosophy specializing in industry and labour, both in teaching and in research.

Although her remarks were addressed mainly, as a candid friend, to the Israeli unions, much of what she said is applicable to the wider union movement, as Patrick Fisher, of the TUC, generously acknowledged.

She was dealing with what she described as "the ambiguities which characterize a modern trade union in its activities and its status within society."

Dr. Bar-Yosef pointed out: "Assuming that the main trend of a modern society is towards changing occupations, changing careers, and the traditional breakdown into classification, into blue collar or white collar, into skills which are formally defined—the image of having one occupation career for the whole of life—if all these are changing, then any organization which strengthens the particular identification with one type of occupation is detrimental to the development of the new occupational pattern. . . . The trade union is often a conservative force."

She wondered if trade unions, in their desire to protect members in disappearing jobs and postpone reorganization, were "happily waiting and not doing anything, and just hoping that the problems would be solved".

Dr. Bar-Yosef visualized situations where unions could make agreements repudiated by their own members, and where modern management techniques could introduce "plant democracy" without the mediation of the trade union.

She was clearly emphasizing the importance of communications within the unions, and obviously the bigger the union the more intricate the maintenance of contact between the leaders and the led.

But one of the most startling ideas in this conference came from Sir Denis Barnes. In his vision of the distant future, he wrote, with disarming casualness, "for various reasons trade unions and employers and their associations may fade away". Mr. Frank Pickford, who deputized for Sir Denis at the conference, commented: "I can't help thinking that Sir Denis had a bit of a twinkle in his eye when he wrote that."

The main purport of Sir Denis's contribution, however, was that organizations on both sides of industry should modernize themselves with a greater sense of urgency than is now apparent.

He suggested that employers' organizations should improve and develop their services to individual firms in the provision of expert advice on such important matters as manpower planning, wages systems, methods of employee participation and negotiation, and the handling of redundancies and training.

He continued: "It is equally the case that employees should be able to turn for support, advice, and help from their trade unions. An effective trade union movement, adaptable to changes in the pattern of employment and nature of jobs, is of vital importance to the solution of our problems."

Sir Denis pointed out that in the last 20 years the growth of trade unionism in Britain has been very slow and is now declining. He explained: "This failure reflects the fact that union organization in Britain is highest in the declining industries and lowest in the expanding industries; and, secondly, that trade unions are not recruiting members quickly enough among the white collar occupations."

He continued: "It is true that since 1948 membership of white-collar unions has increased greatly—more than thirty times greater than the increase among manual workers. This has prompted many people to speak of a boom in white-collar unionism. But this is an illusion. In spite of the phenomenal growth of some white-collar unions, white-collar membership in general has done little more than keep abreast of the increasing white-collar labour force. The importance of manual workers to the trade union movement should not be underestimated. But the expectation is that the number of white-collar employees in Britain is increasing so rapidly that by the 1980's it will exceed the number of manual workers. If trade unions concentrate too much on increasing membership among unorganized manual workers and put too little effort into the white-collar field, they will run the risk of becoming 'the increasingly outdated representatives of a declining industrial minority'."

Sir Denis declared: "If the need for effective and adaptable trade unions is accepted, the trade unions and the TUC face a formidable task. If adequate labour–management co-operation in the handling of technological change is to take place, and if employees are to be represented in this process through their trade unions, then the TUC and the trade union movement must take action quickly. There is need for the TUC and the trade unions to look at their organizations and to bring about changes to promote the development of their membership in accord with the occupational trends in the economy."

He suggested: "Firstly, it seems necessary for the organization of the TUC itself to be flexible and adaptable. This is difficult because the inbuilt balances of power in institutions tend to be in favour of the *status quo*. In the TUC the balance of power is with the manual workers' unions and to some extent, within that area, too much still with the unions in declining industries. This could be one explanation of the relative failure of trade unions to expand as rapidly as they might have done in the growth sectors of the economy. It may also partly be the explanation of the relative weakness of the TUC as compared with trade union centres in other countries, in co-ordinating and guiding the activities of its affiliated unions, adjudicating between their rival claims, and providing them with expert advice. Failure to adapt at the centre can contribute to the failure of individual unions to adapt."

Sir Denis continued: "Unions need to change their concept of negotiation to give more effective protection to their members and to promote their interests more vigorously. They are faced by two major developments as a consequence of technological changes. Firstly, because managements must give more detailed attention to the manpower problems of change, and, for other reasons, much negotiation about the terms and conditions of employment must take place at company or plant level. Generally speaking, however, British trade unions have been accustomed to attach importance to negotiations at the national level of industry and to concentrate their efforts there rather than at local level. National negotiations on some issues will no doubt continue, but unions need to be more effective at company or plant level, and this requires a significant change in structure, organization, and staffing and, in particular, a much higher degree of efficiency and expertise.

"Secondly, because change throws up a new range of manpower problems, negotiations must cover a wider area than the traditional subjects of minimum wage rates, normal hours of work, holidays, and so on. Finally, if the changes which are necessary are not to be delayed and the trade union movement as a whole is not to be weakened in the longer term, the problems of spheres of influence, mergers, and amalgamations must be faced."

Sir Denis concluded: "The TUC and the trade unions are, I think, moving in this direction. There have been more mergers and amalgamations in recent years than for a long time; unions are increasingly engaged in new-style collective bargaining. The TUC and the Confederation of British Industries are reviewing the processes of negotiation in the light of observations made by the Royal Commission. The need is for these and other changes which are taking place to proceed more rapidly."

Mr. David Basnett, an industrial officer of the National Union of General and Municipal Workers and a member of the TUC General Council, promptly pointed to some omissions in the paper prepared by Sir Denis. He had not mentioned, for instance, the increasing degree of state intervention in industrial relations. There have been three significant developments in recent years; there was the National Board for Prices and Incomes, then the Prices and Incomes Act, and now the establishment of the Commission on Industrial Relations.

Mr. Basnett observed: "I think there are two characteristics of British industrial relations which we have inherited from the past. The first is that there has been a complete lack of involvement between management and employees in the objectives of the industry in which they work and the methods by which these objectives are reached. I think there are two reasons for this. One is our history, in which the accepted division of labour removed from workpeople the need for judgement, and the other was the lack of desire of management to involve their employees."

Secondly, it was true that, in general, negotiations in Britain on wages and conditions were conducted on a national level. This meant that the average effect produced was not necessarily appropriate to each plant. Mr. Basnett added: "Nor was this method associated with change in industry, and certainly if you look at the industrial agreements in Great Britain you will find that they provide, compared with America, perhaps very minimum coverage on wages and conditions."

He continued: "Now these two characteristics are changing. I would like to illustrate these changes from an experience in the chemical industry where there is a high rate of technological change. These changes are taking place through the process of productivity bargaining. Here are two such bargains within one company over the last five years.

"In this company, some 5 years ago, we decided that the rate of overtime working was too high. What we did was to produce a very simple proposition. It was that the number of hours to be worked should be reduced by 15 per cent and to ensure that the employees did not suffer any loss of earnings they would get a bonus of 25 per cent but to get this bonus they had to produce work plans by which the amount of production would be precisely what it was previously, despite the 15 per cent cut in time. Now we did not, neither did the management, lay down how this should be done. What we said to our shop stewards was that they should sit down with management, they should involve themselves, and produce the work plans. This is precisely what they did.

"Again, some 12 months after that event, due to some technological changes we had to reduce the labour force by 1 in 5, and we had to increase the effectiveness of those who remained. We discussed, we and the management, together with the workpeople and the managers, and left to the local people the determination of how the objectives should be reached. Again there was an agreement. This is a British situation, a degree of involvement which at that time may have been unique but is today becoming apparent in far larger sections of British industry. This change, this involvement in industry, is becoming more usual."

Mr. Basnett proceeded: "I think that we are beginning to recognize that the type of negotiating machinery we had, with strong national agreements, is beginning to become ineffective. There are a number of industries in which national negotiations have been abandoned. What the national negotiators have done is to convert themselves into advisors to the localities. Indeed, in the chemical industry this has been done, and now negotiations are conducted by some 200 firms with the agreement of the central organizations."

He suggested that these changes are taking place partly because of technological developments and partly by the "happy coincidence" from a rise within managements of a desire to undertake participative management, to involve people. Some trade unions, too, realize that with the satisfaction of their basic economic aims they can extend their influence and undertake a greater degree of joint regulation in industry."

Mr. Basnett conceded, however, that "there has to be a great deal more professionalism on the part of our managements and trade unions in coping with the rapid change in industry. I think the pity is that our institutions have not changed as quickly as people have changed. Our institutions have got to reorientate themselves to service the growth and the application of this professionalism. This is a change one hopes to see both from the CBI and the TUC."

He concluded: "These developments must be conducted against the continuing debate on the extent to which society can allow technological change to determine how it changes, and to the extent to which it controls the changes of technological advance."

Mr. John Peel, the textile workers' leader who is also a member of the TUC General Council, added that the pace at which company bargaining developed in Britain was not determined entirely by the "intelligence or the enlightenment" of union leaders or management, but also by the structure of different industries. In his own wool textile industry there were 1000 firms employing 100,000 people. These firms varied in size and in degree of efficiency and profitability. He said: "For some time we are going to have to hang on to our system of national negotiations. We are developing company bargaining, but only where circumstances allow us to do this."

Mr. Peel acknowledged the importance of mergers of unions. He accepted that some of the criticisms of the TUC by Sir Denis had been given "in a friendly sort of way". But added Mr. Peel, "In some ways I think the TUC is a prisoner of its history. We are too sensitive to our history and, in my opinion, too insensitive about our public image. One of our problems is the maturity of our institutions. This is what is holding us back. The deeper your roots are, the more difficult it is to transplant, to adapt, and to make changes, and while it is true that our maturity in the field of industrial relations is an accumulation of wisdom and experience, it also involves built-in prejudices."

He continued: "There is a current argument going on concerning the role of trade unions in our society. I hope with all my heart that out of that argument comes more influence and more authority for the British TUC to give a lead to its affiliated unions."

Mr. Peel concluded: "I rather think that the trade unions will not fade away but will flower into something new, something wider, and something bigger. And I think the TUC in particular, as well as the movement generally, will become a more professional body, and it will interpret more positively some of the issues forced upon us by technological changes."

Mr. C. Henniker-Heaton, director of the British Spinners' and Doublers' Association, observed: "I've heard of people wanting to abolish trade unions and employers' associations, but I've never heard previously of anyone wanting to abolish them both together! Of course, Sir Denis may have the right solution in this. But he also conceded that we have a very real role to play at the moment."

Mr. Henniker-Heaton recalled the excellent labour relations in the cotton industry without a serious strike for many years, although there had been periods of prosperity and of great depression. He added: "We early agreed to exchange full information with the unions. My technical staff had the authority to give all their facts and figures to the unions and to discuss matters with them. And we have gradually been adopting work study. But we have a saying that 'What Manchester thinks today London thinks tomorrow'. Actually, we are about 5 years ahead of the Royal Commission's report. We sat down for 2 years solidly and we negotiated an agreement which I think takes account both of the need for plant bargaining flexibility and a sensible control. We got together with the unions and, as far as we could, introduced scientific work studies which, of course, are not scientific though we get somewhere near to it."

He pointed out, however, "that there was a danger of plant bargaining being used to bid up wages in an inflationary manner, particularly when industries are concentrated in one area. That is why it is important to retain some form of national bargaining and guidance from the employers' associations and the unions centrally."

Lord Douglass, a former chairman of the TUC, had some reservations, too, about local negotiations. True, the unions could do a great deal of work locally, but it was important to retain some central guidance to prevent local settlements running awry and cutting across each other.

And Mr. Theodore Kheel emphasized that even if negotiating professionals were employed by the unions, it was still essential to have strong and responsible leadership at the head of the unions.

There is obviously a significant difference in negotiating customs in Britain and in the United States. This was brought out clearly by Mr. P. L. Siemiller, president of one of America's most widely scattered unions. He explained: "Collective bargaining is conducted differently in the United States from all other nations. Under our system workers within the same bargaining unit in any given plant have a right by law to select the union that will represent them for bargaining purposes. Once they have done so, that union becomes their sole and exclusive bargaining agent. In most cases, bargaining is carried on at the local level with full participation by business representatives who are locally elected and bargaining committees who are locally elected. Before any agreement can be signed and become binding it must be ratified by the membership as a whole. Thus the mechanics of bargaining are close to the work-force and the results of bargaining must meet with their approval. This is the American system of collective bargaining, and though it is unique, it is the one I must discuss because it is the one I know."

Mr. Siemiller continued: "The purpose of collective bargaining is, of course, to negotiate a contract that will govern the relationship between labour and management in a given plant or company. Such a contract reflects the specific conditions and circumstances prevailing in each particular workplace. It also mirrors the total relationship between a specific employer and his employees. Since no two workplaces are identical in the conditions, circumstances, and relationships that prevail, no two contracts are exactly alike. All, however, have one thing in common.

"They seek to provide ground rules and a mechanism through which the inevitable day-to-day conflicts between labour and management can be adjusted equitably without loss of working time throughout the life of the agreement. To be acceptable in the workplace the total collective bargaining contract must, of course, be acceptable to the work-force. To ensure this acceptability the union leadership must maintain a dialogue with the membership both before and after they negotiate with the employer. This is because, from the union point of view, collective bargaining implies a collective consensus. To the extent that it achieves this consensus collective bargaining humanizes the master–servant relationship. And this is critically important at a time when technological changes threaten to make that relationship even more remote and impersonal than it is already."

Mr. Siemiller declared: "The most important thing that can be said of the American system of collective bargaining is that it has worked. It was devised to ensure that the workers would receive a fair share of the wealth their labour produced. This it has done. But now collective bargaining faces a more significant challenge. Through the processes of collective bargaining, labour and management must seek to work out acceptable formulae for cushioning automation's impact on the work-force. Because collective bargaining requires membership participation, these formulae are not always easy to achieve. In some cases, where a new form of technology threatens to decimate an entire work-force, workers demand, logically, that their union should try to restrict—as well as regulate—the introduction of automated equipment."

He recalled the recent strike of East Coast longshoremen. Here the issue was "containerization", which involved the technical ability to load a container the size of a railroad freight car from a siding to a ship with no more than one man operating a crane. The union leadership did not oppose containerization in theory. They realized that eventually containerization was inevitable.

Mr. Siemiller observed: "At the same time they had to respond to the reality of the situation, and that reality was, of course, that their members were fighting for economic survival. Complete containerization would mean the complete and sudden end of practically all longshoring jobs. Through a long, hard period of negotiations, accompanied by a long hard strike, this conflict was eventually resolved through collective bargaining. In exchange for other concessions the union agreed to a partial introduction of containerization. Undoubtedly the use of containers will be gradually expanded in future negotiations. But this will permit a gradual phasing out—rather than a sudden and abrupt termination of the labour force."

He added: "The lesson of this strike is that while gradual, orderly introduction of technological change is acceptable, wholesale dislocation of an entire work-force is not. Thus, one of the first responsibilities of collective bargaining is to make technological change socially and economically digestible."

Mr. Siemiller stressed, however, that in most cases the role of bargaining had not been to delay change but to find alternatives for those who were directly displaced by automation. This means negotiating for contract provisions that cushion the collision between men and machines. Many unions, he claimed, had made a good start in this direction. For example, the Machinists, the Steel Workers, Communication Workers, West Coast Longshoremen, Mineworkers, Meat Packers, and many others have developed and negotiated contract clauses providing for retraining and relocation of workers displaced by new technology. Collective bargaining has also helped to protect those who, for one reason or another, cannot be retrained. Some of the alternatives include generous severance payments or early retirement with adequate pensions.

During the discussion on his paper, Mr. Siemiller reiterated: "I do not believe that if you are in a free society there is any alternative, any substitute, for collective bargaining. You must have that—and the right to strike—otherwise by government or by decree you have a dictated settlement of a dispute between workers and management, or workers and governments in some cases, and if you have a dictated settlement in one segment of the economy, it is only a short step to having dictated settlements in other segments, and sooner or later, you develop a totalitarian society."

He recognized that there had been misgivings in some places because the unions had negotiated agreements with individual employers providing for the seniority system. This is commonly described in Britain as "last in, first out".

This procedure had been referred to by Mr. Theodore Kheel in a previous debate at the conference, when he explained that if the United States Government deliberately encouraged a policy of creating unemployment as a curb on inflation, the primary victims would be the minority groups who had been last to enter the job market and would be the first to go. Mr. Kheel added: "I think that a large part of the misunderstanding that has developed in our country between trade unions and those in the ghetto stems from an understandable failure to comprehend the significance of seniority to the trade union movement, and what it means to a man who has had a job for 15, 20, or 30 years to be asked to step aside for someone who has lately come into the market."

Obviously, any hint of difficulties between the unions and minority groups, especially Negroes, is both sensitive and explosive, and it was not surprising to hear a passionate defence of union attitudes from Mr. Harry Van Arsdale, who has long been noted for impressive work among electrical operatives and is president of the New York City Central Labour Council. He explained that 200,000 of his union members were Black, "and they are among our best members and we're as proud of them as any members that we have". Mr. Van Arsdale is bitter about some people who have developed what he alleged to be a vested interest in stimulating discord and violence among minority groups.

He declared: "I want to say that some people try to make it appear that organized labour has been responsible for all the discrimination in our country. But we—all the unions and our different organizations—only represent 15 million—for which I am somewhat ashamed—out of 75 million workers in our country. How could we be responsible for all the discrimination, even if we planned it every morning and every night?"

(In Britain, by the way, there are about 10 million trade unionists in a working population of slightly fewer than 25 million.)

But whatever the problems and criticisms of the American trade unions, they have a sturdy defender in Mr. Siemiller. He declaimed: "We must have strong trade unions. Trade unions must have the right by law to act, and they must be responsible. If you have free trade unions, as we have in America, where they have free elections, they are going to assume the responsibility. And I would submit that, in my judgement, the trade unions of America are responsible to their membership; they have a great concern for what happens in America, and they are just as patriotic as any other segment of our society. And they are just like any other human beings."

Mr. Siemiller added: "God created each and every one of us inherently selfish, and as we go through life we don't divest ourselves of very much of that selfishness. Consequently we band together, we do whatever we can to retain as many of the goods, the unit of human wealth, of human labour, whether it is measured by dollars and cents or whether it is pounds or material goods. We use the pressures that groups of people can play one on another in order to provide the best possible society."

Mr. Cyril Plant, of the TUC, inquired if those American workers who had succeeded in securing a shorter working week were now gravitating to be two-job men or enjoying their added leisure.

Mr. Siemiller delighted the conference by his refreshing frankness. He replied: "You hit right at the heart of it. Unfortunately, many people are wanting to maintain a better standard and that means too many are performing two jobs. You take our nation's capital, for example—practically every other taxi-driver will be a government employee, driving a cab in his off-duty hours. These things are happening. We call it moonlighting; I don't know what you call it in Great Britain, but as we copy you, you probably had the term first."

He continued: "There are all too many so-called moonlighters in America today. A good many of our policemen wind up as cab-drivers, and in some cities as bar-tenders. Chicago just put on a drive to make sure that their off-duty policemen were not tending bar in some joint. Not only are government employees participating in two jobs, but so are machinists, and some electricians, and others. It is a problem."

Mr. Siemiller pointed out, however, that because of shorter hours a great leisure industry has grown up and made many new jobs available. He recalled that during World War II, when the War Labour Board could almost order conditions of employment, the unions secured the principle of paid vacations on a wider scale than previously. At first it was one week, but now "we run up to Sabbatical leaves and such things, and 2 weeks, 3 weeks, and even 4 weeks are quite common vacations each year".

He added: "Prior to that time hotels and restaurants were marginal products. Today they have become one of the prime industries because the trade unions demanded leisure time for their membership. While some might practice two jobs part of the year, when vacation time comes they travel extensively. That has helped to create new jobs, distribute wealth, require more automobiles, more congestion on the highways, more policemen to direct the traffic, more hotels, more motels, and more resorts. This process must go on. To create additional new jobs to absorb those displaced by machines, we will probably have to shorten the work day, the work week, and even the work life of a human being more extensively than in the past."

But Professor Ben Roberts, of the London School of Economics, introduced a chastening note. He asked how Mr. Siemiller saw collective bargaining providing solutions to the most pressing problems in the United States today. Professor Roberts explained: "These problems also affect many other advanced countries, including mine, though perhaps not to the same extent as yours."

He declared: "What I see is that collective bargaining has significantly failed to solve the problem of the low-paid worker; significantly failed to solve the problem of poverty; significantly failed to deal with the problem of inflation; significantly failed to provide adequately for the transfer, the mobility, the creation of the new kind of labour force needed to meet the new technologies."

Professor Roberts concluded: "All these problems have had to be met by governments interfering; perhaps you would put it by governments having to do these jobs. As I see it, the kind of system you have extolled, of free collective bargaining, does not seem to have within it the dynamic capacities to meet these challenges. It is only in a newly developing tripartite system, where governments have to play a major role with their appropriate agencies, that we can meet these problems."

Mr. Siemiller protested that he had been obviously misunderstood. It was not his belief, nor had he intended to convey, that collective bargaining itself is the cure for all evils, or the doctor who has the answer to all our ills.

He had declared that collective bargaining can play a role in reducing industrial conflict. He added: "Collective bargaining can also be used, if you please, for other than the workplace; for example, if there is a confrontation between the Blacks, the haves and the have-nots, in the ghettoes; in the community in connection with the schools. But collective bargaining itself does not necessarily produce the jobs. But it protects jobs that are in existence, and protects human beings from being removed from their jobs because of technology without something being provided for the individual for the future. It plays a necessary role in a free society. It provides an opportunity for masses of human beings to have their voices heard at the table where decisions are going to be made."

And a spirited defence of collective bargaining came, too, from Mr. Theodore Kheel, who has a national reputation as an arbitrator in many complex disputes.

Collective bargaining, he insisted, had controlled the opposition to automation. The collective bargaining process is "setting standards for the resolution of group disputes which, though not undertaken wholly by organized labour, are being copied by other groups in our society. We are living in a period of group action. Everybody is getting in on the act. Students are organizing, members of the minority groups are organizing. In America we see football players, baseball players, priests, rabbis, ministers, horse owners, farmers, everyone—all organizing, and what they are doing essentially is copying the practice of organized labour in seeking to achieve first organization and then collective bargaining."

Mr. Kheel added: "What is not thoroughly understood is that in the collective bargaining field we have developed techniques which lead to the resolution of problems. The system can solve problems because it can define them. It is my belief that collective bargaining is, in fact, a better safeguard against inflation than the restrictions that government tends to impose. Government proposals usually provide not only a ceiling but a floor as well, and lead to dislocations in the market. Certain increases are made that are not warranted by productivity. Other increases are not permitted where they could be a further incentive towards automation."

He continued: "One of the facts that stands out in our society is that to a greater degree than there have been wage pressures, has there been a further management response to automate. Automation is going through our society at a fantastically rapid rate, and it is not being opposed either in fact or in practice to any significant extent by organized labour."

Mr. Kheel concluded: "To encourage the use of automation, which is the basic philosophy we bring to this conference, we should discontinue the practice of constantly running down the collective bargaining process. We should see its values for what they are. We should accept collective bargaining as the only viable means by which disputes between organized labour and management can be settled."

But in his final summing-up at the end of the conference, Mr. Kheel conceded: "As for the limitations of collective bargaining; it cannot solve the problems of inflation; it cannot solve the problems of poverty; it cannot solve the problems of the unemployable; it cannot solve the problems of balance of payments. But I don't think its inability to solve these problems and others like it should be called a failure on the part of collective bargaining, which has solved so many problems in other fields. And that, and that only, is my disagreement with Professor Roberts."

Obviously, the values and shortcomings of collective bargaining had been given a healthy airing, as had the differences and similarities of trade union practices and objectives in Great Britain and the United States.

How different in attitude and experiences were the contributions on trade unions from the Israeli participants.

The trade union set-up in Israel is unique. There is no organization in the world quite like Histadrut, the Israel General Federation of Labour. It is a double-headed giant. Most of the eighty-seven men and women who met at Haifa in December 1920, at the first assembly of Jewish workers, were young and all of them were immigrants. They had all come to their historic homeland to get away from the precarious plight of Jews in the countries in which they had been born. They came to be free. They wanted to establish their own homes. But they had to create their own settled life. For few of them had permanent jobs. They changed from day to day, perhaps working one day in a citrus grove and the next on the construction of a road.

But during the passage of the years their organization grew not only in providing security for their members but jobs for them, too, and a pattern for the emerging state.

Today, Histadrut is a federation of thirty-five unions, open to all men and women of 18 and upwards "earning their livelihood by their own labour without exploiting that of others". There is no discrimination against race or faith. And in a country with a population of fewer than 3 million, Histadrut unions have a remarkable adult membership of 970,000.

But Histadrut is also a vast trading and commercial organization. Its concerns produce two-thirds of Israel's farm produce, 40 per cent of the country's building activity, and its affiliated enterprises contribute more than 25 per cent of the total national industrial output. It has its own bank, insurance company, and several specialized subsidiary organizations which contribute significantly to the financial basis of the whole structure of the labour economy. Next to the government, with which it maintains close ties, it is, in effect, the Second Estate of the country. It has ramifications over a wide field of industry, being involved in metal, cement, glass, ceramics, plastics and chemicals, rubber, timber products, and electrical equipment.

Not surprisingly, there are strains and stresses arising from these multifareous interests, and Israeli participants were frank in disclosing some of them to the conference.

Colonel Israel Meidan, the director of the Israel Institute of Productivity, was impressively candid. With the continued growth of the economy, he pointed out, social values have developed, too. "These changing circumstances", Colonel Meidan declared, "have created areas of tension, and the Histadrut has had to re-evaluate certain basic concepts. Because of its relative power position, the Histadrut has found itself without any countervailing balance. Dominating, as it does, entire sections of the economy, the Histadrut has had to exert self-discipline in restraining its demands on the economy as a whole. The position of the Histadrut, and the nature of its dominant social orientation, has led it to be a direct partner in other areas of national economy. It is represented at various levels of economic planning and, in particular, participates in the sectoral committees which include representatives of management, employers, government, workers, and professional specialists. These committees have the responsibility of evaluating future trends in product development, competitive capacity with respect to imports, manpower development, and investment policies. The development of Israeli industry, and the growing complexity of the economic system, has created the realization on the part of the Histadrut that it must become a partner in stimulating technological development."

Colonel Meidan continued: "This central position of the trade union movement at the national level, and the acceptance of social responsibility that this involves, has created a dilemma for the unions. Histadrut leadership has more than once found itself in conflict with its own rank and file over such questions as wage stability, labour mobility, and manpower requirements. This duality of interests may find expression in many instances."

He explained: "Although in favour of labour mobility at the national level, Histadrut finds itself supporting such principles as security of tenure, which restricts the transfer of redundant personnel while protecting the less efficient worker. A further case may be when union leadership emphasizes the need for rational work measurement and finds itself opposed by workers at plant level interested in maintaining the advantages of the *status quo*."

This dilemma, Colonel Meidan pointed out, becomes even more acute considering the political structure of the trade union institutions. The high degree of centralization in the trade unions does not make the problem less difficult. This situation tends to hamper the maintenance of trust and communication between rank-and-file members and union leaders, and has led to "wild-cat" strikes.

He continued: "In spite of these conflicts, the Histadrut's participation at the higher level contributed significantly to industrial progress and the acceleration of technological change. The Israeli worker has increased his productivity by 5 to 6 per cent each year for more than a decade now. Although there are some instances of maintaining redundant labour, in no case was there opposition to the introduction of new equipment or processes."

Colonel Meidan advised: "Future technological developments, the need for greater specialization, and acceleration of the pace of change will require more extensive and effective communication between the leadership and rank and file of the trade unions. The existing conflict between the responsible long-range economic outlook and the immediate demands of the workers at the enterprise level might become more acute. To alleviate this difficulty the unions will have to extend training and develop more extensive programmes designed to discuss with the workers the implications of the economic scene."

It was also reported by Colonel Meidan that industrialization has developed greatly in the *kibbutzim* in recent years, and although there are great differences in the quality of leadership in these co-operatives, in some there is a high level of managerial skill and the increase in productivity in this sector is higher than in any other in Israel.

But in some other enterprises of the Histadrut there is little sense of participation by the employees, and there are few cases of workers' representatives being seconded to managerial functions. In one of Histadrut's oldest plants—Phoenicia Glass—

an attempt to stimulate participation to express a sense of partnership provoked opposition from both workers and management. The workers suspected the step as an attempt to manipulative control, and the management feared the loss of decision-making authority to inexperienced personnel. Colonel Meidan commented: "It is clear that merely bringing workers' representatives into managerial functions did not increase participation. Communication gaps arose and the lack of careful preparation produced adverse results rather than those expected."

Nevertheless, in Israel generally, workers' shop committees are an accepted element of consultation, and joint productivity councils are a frequent feature of the industrial relations system, and have generally proved to be advantageous.

Mr. Joseph Hausmann, who is a director of the labour department of the Manufacturers' Association of Israel, also referred to some of Histadrut's difficulties. Its close association with the government, he said, has led to confrontation with its own members. To minimize these confrontations and repay its obligation to its centralized power, the Histradrut, though a voluntary organization, will either have to rely on a law or improve the trade union education of its members.

Such a law, Mr. Hausmann suggested, would provide a set of adequate penalties for breaches of collective agreements, and provide rules to regulate collective bargaining. But by opposing such legislation, yet approving regulations to render strikes in the civil service and essential supplies illegal, the Histradrut has found itself "open to the censorship of public opinion".

Mr Hausmann acknowledged that the intricate problem of participation gets more involved the higher its level. The stronger the institutionalization of participation, the longer and weaker the channels of communication between the representatives and the represented.

There was a demur from Mr. Chaim Kohn, the director of Histradrut's productivity division, who thought Colonel Meidan had overstressed the conflict between Histadrut and its trade union members.

Mr. Kohn observed: "The phenomenon of a worker, or a group of workers, not always agreeing with the policy of their trade union is universal, and perhaps natural, particularly in the field of increasing productivity and technological development. It sometimes happens that an individual or a group find themselves unfavourably affected by some activity on the part of management. So it seems to me that the real problem here is the conflict of the worker as shown by immediate results and the long-range attitude. It is the duty of Histadrut, or any trade union, to explain developments to the workers from the long-range approach, so that they should understand that a temporary recession in the status of an individual or group can assist the entire population, including themselves, to advance and raise its standard of living. I think we have reached a point where the majority of workers find no contradiction between Histadrut action in the field of increased productivity and technological advancement and their own real interests."

But Mr. Kohn emphasized that improved participation from workers depended first and foremost on the genuine goodwill of management. He added: "The role of management in this field is decisive."

Mr. Theodore Kheel commented sympathetically on the dilemma of a workers' representative participating at the highest level of a company. He explained: "The representative must be persuasive with the people he represents, but he must also be persuasive with the people with whom he negotiates. If he is too much of a statesman there are wild-cat strikes: if he is too responsive to the rank and file, he is irresponsible."

And Mr. Fred W. Cartwright, of the British Steel Corporation, recalled that when the steel industry was renationalized in Britain, three union leaders were appointed to the board. They had to give up union office. The management kept back no secrets from them. But here was the problem: were these new-type directors to reflect the views of the shop-floor to management, or the view of management to the shop floor? But they were no longer in the inner council of the union, so they do not know what the union view is, and as they are not in the executive part of management, they are not really on the inside lines of production. They feel rather frustrated. Mr. Cartwright confessed: "I am not sure whether working directors are a good thing or a bad thing."

But Mr. Dai Davies, the British steelmen's leader, had no doubt. He admitted: "I have always been old-fashioned enough to believe that it is the function of management to manage, and it is vital for the trade unions to retain their independence, to be free either to support or criticize management."

Regarding the Steel Corporation experiment with three ex-union directors, Mr. Davies was emphatic. He declared: "Although it is only an experiment, due to be reviewed in a year or two, I am bound to say that I never regarded the exercise as anything more than a well-intentioned gimmick—but a gimmick nevertheless."

A far wider sweep of the role of the trade unions was introduced by Mr. Yeruham Meshel, an executive member of Histadrut.

He submitted that it is difficult to see today how trade unions can preserve their status as legitimate representatives and champions of the social claims of the workers without active participation in the creation of economic conditions leading to the rise in the standard of living of the working community.

The Swedish trade unions, he pointed out, have come to this conclusion in a very methodical way, and it seems that trade unions in other countries are beginning to follow in their wake.

Mr. Meshel added: "It should be stressed that this approach does not contradict the principle that the strike weapon as an *ultima ratio* for union struggles must remain in the hands of the trade unions as before. In the consciousness of the workers, the right to strike has become one of the basic rights, and all those favouring the system of an enforced solution, such as compulsory arbitration or the complete banning of strikes in the 'vital services' (which, in fact, means all public services and many others) must encounter some reasonable resistance by the unions."

But, Mr. Meshel proceeded, "However, logic would suggest that this positive attitude towards the progress of the national economy requires the maintenance of normal labour relations, strict fulfilment of collective agreements, and abstention from using the strike weapon as long as there are other suitable means for settling disputes. An aspiration towards industrial peace is part of the social duty of the trade unions."

He continued: "I have to emphasize that settling the differences between groups of workers in different occupations, and finding a common denominator between them is a problem no less acute than the encounter between the unions and employers. The sharp contentions between the various workers' groups, or within the unions themselves, prove more clearly our premise concerning the need for active participation by the unions in the development of the national economy. If it is true that a united union movement is a stabilizing factor in the labour relations system of the national economy, then the unification of the various groups of workers under one general administrative framework is in itself a social task of exceeding importance for the trade unions. It is quite reasonable that the social aim of bringing together different groups of workers in united trade unions should be of particular importance in the free democratic countries which believe in the central planning of the country's economy."

He included in this list France, England, Italy, Holland, Belgium, Austria, Sweden, Norway, and Israel.

Mr. Meshel declared: "Planning the economy without the consent of the free trade union is absolutely out of the question today. In addition, it is very doubtful whether it would achieve any reasonable success without their active support for the planners' intentions. To differentiate between the social and economic problems of the unions is possible only in the abstract; in practical life they actually constitute one fabric and are almost inseparable."

There was impressive agreement among practically all the participants that trade unions must be an essential partner in the advancement of technology. The impact of this technology on collective bargaining will be massive, and inevitably governments are bound to become increasingly concerned in industrial relations.

Indeed, Professor Roberts thinks there will be larger roles for governments in this field than in the past.

He explained: "Our traditional system of industrial relations has been based on the view that there should be a minimum role for government, either through the law or governmental agencies. My own view is that we have been moving steadily over the last 50 years, not at an even pace, towards a greater role for government. It is perfectly clear in the type of legislation that the present British government is seeking to bring in, that it will exert a greater influence. We are moving from a two-party system in industrial relations into what I would regard as a tripartite system. This does not mean that the government should seek to take over the functions of employers and unions in settling terms and conditions of employment, but to provide such conditions, and exercise such influence, as will enable unions and employers to arrive at settlements of issues in conflict, or even issues not in conflict but in the interests of the community as a whole."

Professor Roberts concluded: "This, I think, is the philosophy towards which we are proceeding. The argument is about where the lines should be drawn. We have not yet settled down into a new pattern of relationships, but this we shall do over the next decade or so. If we look back, say, from the end of this century, we will, in my opinion, observe the evolution of a new set of relationships. And I think the government might attempt to raise the standards of negotiators. It should seek to give the parties whatever assistance it can. There may be a variety of different means to promote more efficient management, more efficient trade unionism, so that settlements can be better related to the needs of the community as a whole."

Professor Jay Tabb, of Technion, had one qualification on Professor Roberts's analysis. He doubted whether governments' intervention needed to increase. After all, he recalled, governments had always been ready to intervene when they found it necessary. He thought the most probable future change would be the increased complication of collective bargaining in an era of automation. Professor Tabb thinks the unions and employers' associations will have a more difficult role to play and will need better trained personnel.

Mr. Harry Van Arsdale (New York), suggested governments could participate by insisting on the continual training of negotiators on both sides, and governments could investigate, when there was a breakdown, which side had contributed the more to the breakdown. Such studies would have to be undertaken only by impartial people who had the confidence of both sides. And a government agency should give credit to participants in successful negotiations; such recognition might stimulate both sides to find a settlement. But the government should not try to impose its own decision, because forcing one side or the other to accept an unsatisfactory result would not contribute to the success of such a contract.

In his summing up, Mr. Kheel threw out an intriguing proposal for future consideration.

It was that, as a *quid pro quo* for the introduction of technological change a guaranteed annual wage might be sought.

Mr. Kheel confessed: "This has come up many times in American industry, and it tends to send shivers down the backs of employers because they anticipate having to assume an obligation which, if economic conditions turn bad, they will not be able to meet. It is, in fact, as they see it, the assumption of an incalculable risk, and one they prefer not to bet on."

He proceeded to argue, however, that "through collective bargaining it has become possible to define the limits of the obligation that an employer is required to assume. There is, first of all, the unemployment compensation which employees who are laid off would otherwise receive. It has been ruled in our country that such compensation can be credited towards an annual guarantee. There are also the vacation, the holiday periods, and so forth which likewise make up part of the guarantee. In addition, there is the incentive for the employer to regularize his employment by changes in the operations of his company. And there is the corresponding obligation on the part of the unions to make it possible for him to regularize employment, by altering vacation periods, by changing hours of work, and by taking other steps that make it possible for an employer to have a more even flow of work."

Obviously, Mr. Kheel is convinced that a guaranteed annual wage has become more feasible because of improved methods of collective bargaining and the advent of technology.

It was the final thought tossed into the conference.

But Mr. Kheel's words are unlikely to be the last on the subject.

Chapter 6

How the World's Poorest Countries Can Be Modernized

THERE is no simple way of bringing the world's underdeveloped countries into the twentieth century. There is not even a simple way of defining an underdeveloped country, nor has any one continent the monopoly of backward countries. Even the Americas can produce examples of communities which are indifferent to the recipes of success so prodigally exemplified in more sophisticated nations. Yet the prospects of progress for advanced countries offered by technological developments make the disparity with underdeveloped countries the more marked and the more embarrassing in a world shrinking visibly through the advance of modern communication and transport systems.

It was, therefore, with some courage that Professor Nicholas Kaldor, of Cambridge, presented a paper on "The role of modern technology in raising the economic standards of the less developed countries". Few people are better qualified to produce more authoritative economic blue-prints. In addition to having been the main economic adviser to British Chancellors of the Exchequer, he has, as Lord Carron reminded the conference, been an adviser "at the highest level to a vast number of countries in the world". Even so, there was no lack of dissent from some of the suggestions made by Professor Kaldor, but he seemed to expect this, and, in fact clearly enjoyed the stimulation of discussion.

Professor Kaldor acknowledged that there was no simple, general answer to the two main aspects of his theme.

The first was the wider one related to the strategy of economic development. Given the limitation of available resources, how should they be applied as between the different economic sectors? Should they concentrate on developing technological standards in the traditional economic activities—which include improving the efficiency of agriculture—or should they concentrate on developing new activities, that is to say, in establishing modern industries?

The second—narrower—aspect relates to the question: What are the most appropriate techniques for adoption in any one industry? Should the underdeveloped countries "copy" the technologies of the advanced countries or should they develop new technologies appropriate to their circumstances? Should they aim at creating maximum increases in new employment opportunities or should they aim at a higher productivity per worker, even if this involves a slower increase in the volume of employment?

Professor Kaldor explained: "The most that one can expect from economic analysis is to set out the relevant factors and establish certain general criteria for judgement."

He proceeded: "It has been a frequent criticism of the post-war development plans of many underdeveloped countries that they concentrated on industrial development on a broad front to the neglect of agricultural development; and that they concentrated on developing substitutes for imports, even when this involved a very wasteful use of resources, to the neglect of developing exports. These "inward-looking" strategies of development meant that despite very large investments of capital—financed by external aid or through varying forms of taxation of the agricultural sector—they bought a very meagre social return, either in terms of additional employment or in terms of improvement of living standards. As a criticism of the post-war economic policies of many underdeveloped countries this is no doubt well founded."

No one who has examined the disappointing record of a number of underdeveloped countries in Latin America or in

South-east Asia, he said, could fail to be impressed by the fact that the pursuit of indiscriminate "import substitution", without regard to cost, led to perpetual balance of payments problems, inflation, low rates of growth, and meagre results in terms of additional employment or higher consumption per head of population. Yet it is a criticism concerning methods, rather than aims; of tactics rather than of long-run strategy.

Professor Kaldor declared: "There can be little doubt that the kind of economic growth which involves the spread of modern technology, and which eventuates in high real income per head, is inconceivable without industrialization. In that broad sense there are no alternative roads to economic development. It is no accident that all 'advanced' countries with relatively high incomes per head have a large manufacturing industry, and in most cases, are also large exporters of manufactured goods."

He continued: "The reason for this is that as real income rises, only a diminishing proportion of income is spent on food and a growing proportion is spent on industrial goods and on services. One could, in theory, conceive of a country 'specializing' entirely in agriculture and obtaining all its industrial requirements from abroad. But it could never absorb more than a fraction of the working population on the available land. Though in all underdeveloped countries the greater part of the working population is 'occupied' in agriculture, most of this represents disguised unemployment; a rural community maintains all its members and expects everyone to share in the work. Much the greater part of this labour could be withdrawn from agriculture if alternative employment opportunities were available without any adverse effect, and probably with a beneficial effect, on total agricultural output. For the relief of the pressure of labour on the land is itself a most potent factor in inducing improvements in technology which raise yields per acre, as well as the yield per man. These improvements normally require an increase in the capital employed on the land; but the savings necessary for the increase in capital are themselves a byproduct of reduced population pressure. The reduction in the agricultural population, and the increased use of capital in agriculture are thus different aspects of the same process."

Professor Kaldor explained: "As there are fewer mouths to feed, the 'agricultural surplus' rises (the excess of agricultural production over the self-consumption of the farming population); the rise in the 'surplus' enables farmers to plough back a higher proportion of their output—in the form of better tools, improved seeds, fertilizers, etc.—and such improvements tend to be both 'labour saving' and 'land saving'; they diminish the labour requirements at the same time as they increase the yield of the land."

Hence a technological revolution in agriculture leading to a faster growth in output is generally associated with a steady reduction, not an increase, in the agricultural labour force. The best proof of this is to be found in the fact that those advanced high-income countries which have "specialized" in exporting agricultural products and importing manufactured goods (and which export the greater part of their agricultural output) such as Australia, New Zealand, and Denmark, have nevertheless only a low proportion of their labour force in agriculture—of the order of 10–20 per cent as against the 70–80 per cent in low income countries who barely produce enough to feed their populations on a very low standard. And it is one of the best-established generalizations of economic history that with improvements in technology, and a rise in real income per head, there is a continued reduction in the proportion of the labour force employed in agriculture. The advance in agricultural technology thus depends on and is conditioned by the growth of the agricultural surplus. At the same time, the growth of this surplus is the main factor determining the growth of employment opportunities in the non-agricultural sectors of the economy.

Professor Kaldor declared: "This has two aspects. In the first place, the growth of the non-agricultural employment-potential depends on the rate of growth of *marketed* as against self-consumed food supplies. Food is the 'wage good' *par excellence*, and any attempt to increase urban employment at a faster rate than the agricultural surplus permits is bound, sooner or later, to be vitiated through violent inflation. In the second place, the growth of the agricultural surplus is an essential condition for providing the growth of purchasing power necessary for sustaining the growth of demand for industrial products. The growth of industrial activities is generally regarded as being conditioned by the rate of capital accumulation in industry. No doubt the accumulation of capital is an essential part of this process; in capitalistic economics, however, the process of industrial capital accumulation is very largely self-generated—it is both motivated and financed by the re-investment of business profits. Hence the rate of capital accumulation is itself conditioned by the growth of profits associated with the process of expansion, and with the growth of profitable investment opportunities provided by the increase of effective demand for industrial products."

In any sustained process of economic growth, he argued, the expansion of the agricultural surplus provides the source of demand for the growth of industry; the growth of industry,with its manifold associated "tertiary" activities, provides the source for the growth of urban employment; and the latter provides the resources, and the incentives, for technological improvements in agriculture, which in turn, by raising the agricultural surplus, ensure the continued demand for industrial goods. Thus industrialization and the growth of agricultural productivity go hand in hand, and are complementary to each other. Any improvement in either one of these sectors facilitates the growth of the other, in the manner of a continued chain-reaction.

This has undoubtedly been true of all the present "developed" countries who have only attained their privileged position—with levels of real income per head that are twenty to thirty times as high as in the underdeveloped countries—as a result of a long-sustained process of rapid technological improvement in both manufacturing and agriculture. In the countries which failed to participate in this process of long-sustained growth, some constraint must have inhibited the process of continued "chain reaction". For one reason or another, the "environment" was not favourable, or not sufficiently favourable, for this process to get really started, or if it did get started, to be long sustained.

Professor Kaldor observed: "I should like to distinguish between four different factors which could provide the cause of such constraints; but this list is not intended to be exhaustive, and in any particular historical case, there may have been other (largely political or educational) factors at work."

(1) The first of these is the lack of responsiveness in agriculture to outside stimuli on account of economic or social factors. In countries in which agricultural over-population has reached very high levels, and where, in consequence, the rural community is very poor, the response to better market opportunities may well be nigh non-existent. Harvests may vary with weather conditions; but a good harvest simply means less starvation; it does not create a higher cash income, nor provide the opportunities for improving the land through more investment. Even when the physical conditions are more favourable, and the population pressure is not so large, ancient forms of land tenure, the survival of feudal institutions, absentee landlords, may make it impossible for the actual cultivators of the soil to exploit the opportunities for improvement in the arts of cultivation.

(2) A second factor, equally important, is the failure to develop exports *pari passu* with the growth of the economy. For the early industrializers who were able to supply the world with the products of advanced technology, the rate of industrialization was "export-led" from the start. For the late developers this presented a far more difficult problem—since they were trading the advantages of low wages against the superior technology of the older industrial countries—but those which succeeded in passing through the critical phases to become high-income countries were invariably the countries which managed to sustain a buoyant growth of their exports.

For many of the countries outside Europe the initial stimulus to development came from the growth of exports of plantation agriculture and mining. With the growth of industry in Europe, markets were created for temperate foodstuffs, tropical products, and minerals. The exploitation of these opportunities sometimes proceeded on the initiative of native producers, but more frequently through European capital and enterprise. It is often contended that the growth of foreign-controlled mines and plantations was a "foreign enclave" which contributed little to the economic growth of the countries in which these developments occurred. This is broadly true, but not wholly accurate. It brought with it the educational stimulus of foreign contracts; and what is more important, it was a source of export earnings which could be channelled, in a suitable political environment, to provide opportunities for developing local industries. By contrast, the countries which did not have resources for the development of exports of primary products—on account of unfavourable climate or lack of minerals in their territory—were under the greatest handicap to get economic development started.

(3) Export earnings through agriculture or mining do not, however, in themselves suffice to get the process of industrialization going without state support in the form of subsidies to industry or the adoption of protective tariffs. In the absence of these, the high initial costs, in terms of agricultural products, of home produced manufactures imposes too severe a handicap on any latecomer to industrialization to make manufacturing activities commercially profitable. The advantage of any underdeveloped country in the industrial field resides in low wages. In the initial stages of industrialization this advantage is more than offset by low productivity. Hence under conditions of free trade, when the domestic price of manufacturers in terms of primary products is determined by world prices, the process of industrialization may never get started.

(4) However, whilst the competition of more advanced countries may prevent the domestic establishment of industries in the absence of *some* measure of protection, such policies are likely to succeed only if they are applied *sensibly* and with *moderation.* Excessive or indiscriminate protection may itself inhibit continued development, for a number of reasons. In the first place, any degree of protection involves a levy on agricultural producers—

who are forced to sell their produce on less favourable terms in relation to the industrial goods for which they are exchanged; if the protection is excessive, the terms of trade will deteriorate so much as to deprive the farming community of the advantage of a growing real income which is indispensable for the adoption of more advanced technologies. Hence it makes for agricultural stagnation which will sooner or later bring industrial expansion to a standstill.

Import duties are efficacious in promoting industrialization so long as there is scope for creating an internal demand for home-produced manufactured goods through the replacement of the pre-existing imports of such goods.

There are dangers in the "inward strategy" of following the creation of comprehensive industrialization on a broad scale, with each separate industry obtaining the differential subsidy required for domestic manufacture, regardless of comparative cost. These are well illustrated in the history of many Latin-American countries. Chile, Argentina, and Brazil, for instance, each passed through a phase of relatively rapid growth following the establishment of highly protective tariffs or important prohibitions during the Great Depression. But in each case this phase was followed by a prolonged period of very slow growth, or stagnation, combined with prolonged and violent inflation, as the growth of urban employment tended to rise faster than the growth of marketed food supplies.

By contrast the histories of many smaller European countries, which have been comparative latecomers in industrialization—such as Switzerland or the Scandinavian countries—show the advantages of an "outward strategy". In each case tariffs were kept low and reasonably uniform; this enabled specialization in industrial development from the start; some industries were developed to the stage of becoming internationally competitive before starting with the establishment of others; the range of domestic production of manufactured goods was broadened only gradually. In this manner the growth of exports kept pace with the growth of production, which provided the means for a steady growth in the employment potential of the industrial sector, and thus made it possible for the disguised unemployment in agriculture to be gradually liquidated.

Professor Kaldor posed the question: What kind of "technologies" should the underdeveloped countries adopt in order to secure the fastest rates of growth?

He pointed out that there has been a great deal of criticism of the policies of underdeveloped countries wishing to "imitate" the advanced technologies of the Western industrial countries without inquiring whether these were appropriate to their circumstances. The advanced countries employ techniques which require a very large amount of capital per worker; since the underdeveloped countries have plenty of labour and very little capital, they should, according to this view, develop "intermediate technologies" which have a lower capital requirement and a higher labour requirement per unit of output. Allied to this is the suggestion that since the great majority of their population live in rural areas, they should avoid the high economic and social cost of large urban conglomerations and, instead, bring industrialization to the villages. The development of "cottage industries" was at one stage an important feature of Indian economic planning; some years later, the idea of creating small-scale metal smelting and engineering works in the villages was the main feature of the "great leap forward" phase of China. Both of these proved a costly failure.

The fact that in all known historical cases the development of manufacturing industries was closely associated with urbanization must have deep-seated causes which are unlikely to be rendered inoperative by the invention of some new technology.

There are very important advantages, Professor Kaldor continued, in the regional concentration of industries which go well beyond the economies of large-scale operations. They are to be found in the availability of manifold specialized skills, know-how, and easy markets which make it profitable for firms using similar or related processes to be located close to each other. The concentration of particular phases of cotton textile production in particular cities (and sometimes in special regions of cities) in Lancashire was not the result of any deliberate planning; it emerged spontaneously as a result of market forces.

The same is true of the concentration of the engineering industry in Birmingham, or of the jewellery industry in the little town of Pferzheim in Germany (which at one time had 20,000 jewellery workers). One could mention innumerable other examples. In none of these cases was the geographic concentration to be explained by the technical need for particularly large-scale plant. The advantages of geographic concentration lay in the opportunities for a higher degree of specialization between different enterprises and the consequent subdivision of industrial processes; in the availability of labour of many specialized skills; in a wide range of engineering and marketing knowledge.

Professor Kaldor insisted that the idea that underdeveloped countries would stand to gain from the use of special "labour intensive" technologies is of doubtful validity. It is, of course, obvious, he claimed, that countries which are short of capital should use techniques which make the best use of capital—i.e. which have a *low* investment requirement per unit of output. It is a mistake to believe, however, that more primitive or less mechanized techniques which require less capital per worker are also more economical in capital per unit of output. A lower capital/labour ratio does not necessarily mean a lower capital/output ratio—indeed the reverse is often the case.

The countries with the most highly mechanized industries, such as the United States, do not require a higher ratio of capital to output. The capital/output ratio in the United States has been falling over the last 50 years while the capital/labour ratio has been steadily rising; and it is lower in the United States today than in the manufacturing industries of many underdeveloped countries. The technological progress in the present century led to a vast increase in the productivity of labour but without any associated reduction in the productivity of capital investment.

One reason for this, Professor Kaldor suggested, is that the greater speed of "through-put" with modern technical processes has led to economies in the amount of working capital required per unit of output. It was found, for example, in India that non-mechanical processes (like hand-spinning or weaving) are more costly in capital than modern machine processes, despite the fact that the cost of a hand-spinning wheel or a handloom per worker is only a small fraction of the cost of power-driven machinery per worker. The reason for this was found to be the very much greater requirement for working capital in the case of hand-made processes. Just because processing takes much longer with non-mechanical methods, the working capital locked up per unit of raw-material input, or per unit of final output, is much greater.

Nor is it correct to suggest, he continued, that just because there is a large amount of surplus labour, labour ought to be regarded as a "free good". However great the level of unemployment, open or disguised, employed labour still has to be paid wages. A higher level of employment means a higher wage bill and a correspondingly higher demand for consumption goods, particularly for food. This is not only because at a low standard of living workers prefer to consume any additional income. Employment requires a greater expenditure of physical energy than idleness; a man can only perform a full day's work with a much higher calorie-consumption than is necessary to sustain life in the absence of work.

For this reason it is often profitable for employers to pay higher wages than the minimum at which they can hire workers; when earnings are low, the calorie consumption becomes insufficient for efficient work performance—hence, beyond a point, a lower wage worker involves a higher wage outlay per unit of output. This, I believe, commented Professor Kaldor, is the main reason why even in the least-developed countries with the largest ratio of surplus labour, the wages of factory workers are so much higher than the "opportunity cost" of labour, i.e. the level of earnings per head in the rural sector.

Professor Kaldor observed: "All this does not mean that the use of the most highly mechanized technologies, yielding the highest output per worker, is necessarily the best means to be chosen. Nor is it correct to suppose that the 'best' technology for underdeveloped countries is that which yields the highest output per unit of investment (irrespective of labour requirements) as would be the case if labour could really be treated as a 'free good'.

"The best general rule, as several economists have pointed out, is to choose the techniques which, at the prevailing and expected level of wages, yield the highest rate of profit per unit of investment. For as a general rule, the techniques which yield the highest rate of profit serve to maximize the ratio of output to the additional consumption generated in producing that output; the more of the additional output is available for re-investment, the higher the rate of growth of capital accumulation which the economy can sustain."

He continued: "The most profitable technology for a capitalist employer may thus be no different from the theoretical 'optimal' technique in a planned socialist economy. The question which needs further examination is whether this 'most profitable' technique is necessarily different in underdeveloped countries with relatively low wages per head, from the 'optimal' techniques in the advanced countries with high wages per worker. If the above characterization of technological progress is correct, and the 'latest' technologies normally yield the highest output per worker without any greater outlay of capital per unit of output, then the latest techniques must be superior to all others, irrespective of the amount of capital required per worker."

But Professor Kaldor suggested that there are two main reasons why machinery of the "latest type" may not be the most suitable for use in underdeveloped countries.

The first is the opportunity to buy, at a relatively low cost, second-hand machinery which, though physically perfectly fit for use, is no longer profitable to operate in the advanced countries owing to the competition of more recent and more technologically advanced plant and machinery.

He added: "It is well known that equipment in advanced industrial countries is withdrawn from use long before it is physically worn out on account of the continued rise in wages in relation to product prices, which causes rapid obsolescence. This gradually eliminates the profit on their operation long before the physical efficiency of the equipment is impaired on account of wear and tear. Such 'obsolescent' equipment can be acquired at low prices; and in countries where wages are low relatively to product prices they may thus be *more* profitable to install than the latest 'new' equipment; and given the opportunity to buy such machinery at low prices relative to their original cost of production, it may well 'pay' underdeveloped countries to adopt, not the latest technologies of the advanced countries, but the technologies which were the most advanced ten or fifteen years ago."

Professor Kaldor continued: "The second reason why the most advanced technologies may not be the most suitable is because they require too large an output for their optimal utilization. Technological advance is biased in favour of large-scale production—the most efficient generating station, the most efficient steel mill, the most efficient tanker, and so on, require a much larger rate of throughput than a generation ago. Twenty years ago the optimal size of an integrated steel plant was around 2 million tons a year. At present it is around 6 million tons, and in another 10–20 years it is likely to be twice as much again. This is partly because there are always inherent economies in larger-scale operations; and technical advance consists in overcoming the constructional problems of making things to an ever larger size. For example, the largest tanker that can be built is necessarily the cheapest, since the carrying capacity of a larger tanker increased more in proportion to the labour and material cost of its construction or the labour and fuel cost of its operation.

"It is partly also because much of the research and development expenditure in advanced countries is undertaken by giant firms, whose economic interest is to develop techniques which give a differential advantage to the large-scale producer. The underdeveloped countries whose market for industrial goods is limited, have neither the resources nor the market opportunities for the installation of plant with a very large capacity. Perhaps the most important respect in which they require differing technologies from those of the advanced countries—and for the sake of which they would be justified in incurring heavy research and development expenditures of their own—is to develop ways in which modern technological processes in steel-making, chemicals, etc., could be adapted to efficient operation on a smaller scale."

Professor Kaldor concluded: "Finally, one cannot emphasize too strongly that the efficiency and speed with which modern technology can be introduced in underdeveloped countries is very much a matter of the quality and character of the system of education. In many underdeveloped countries too much of the resources are spent in providing high level education of a theoretical kind and too little in creating a cadre of qualified managers, engineers, and technicians. Together with the creation of efficient transport and communication systems, this is the most important aspect in which the development of a social 'infra-structure' promotes economic development."

The first comment on Professor Kaldor's paper came from Mr. C. Henniker-Heaton, the director of the British Spinners' and Doublers' Association, and the British employers' delegate to the International Labour Organization.

While he agreed with Professor Kaldor's main conclusions, he had what he called some "minor criticisms".

"It is entirely true", he stated, "that no country can rely entirely on its agricultural production, however export-orientated it may be. But it must be remembered that agricultural and horticultural methods are now being developed which raise these activities to the level of a scientific industry, even if the countryside loses some of its charm in the process. Using these methods, together with natural resources of climate and soil, agriculture and horticulture can still play an important part in getting certain underdeveloped countries off the ground export-wise."

Mr. Henniker-Heaton added: "Then I would like to take a further look at the suggestion that second-hand machinery or plant in good condition, and no longer required by highly industrialized countries, may often be a good investment for developing countries. Now this can sometimes be true. I'm not challenging it at all, but it is a very dangerous doctrine. Experience as well as instinct tell me that it is seldom worth while in the long run. Often the machinery, so very willingly off-loaded by the previous users and offered for sale by eager intermediaries, is not in such good condition or so useful as it appears to be. In any case, such a policy, if carried too far, may mean that the country concerned will stay permanently behind the band as to world levels of efficiency."

He likened a developing country to a large family of a father, mother, ten sons, and ten daughters.

Mr. Henniker-Heaton explained: "The family is past the stage when eight of the sons were engaged in tilling the land or tending the flocks, and when one of the other two was a priest and the other a clerk. Let us suppose, also, that they have got over the stage of civil war among themselves, and they have accepted the direction of their father—in effect, their government. There are still too many of the sons, say four or five out of the ten, engaged in agriculture and mining. Only one is engaged in industry and construction, one in handicraft, two in administration and education, and one is still a priest."

He continued: "These 'manpower distribution figures' are typical of what is no longer an undeveloped country, but is a less-developed or developing country, and it must now develop in the right way and become able to hold its own in a modern competitive context. That is where the family ought to follow the advice of Professor Kaldor, and that is what they would do if they were automata, endowed with first-class electronic brains. But, of course, they don't because they are flesh and blood, and the brains they do have, and particularly of the father or government, are filled with all sorts of things such as tradition, religion, undue hatred of neighbouring families, undue love of your own family, pride, inferiority complexes, instincts, emotions—family pressures—everything that goes to make a fallible human being.

"So, time and again, the father of the family makes a wrong decision, often, as in India, out of the very best of motives. Nor would the position be much better if he formed a democracy out of his sons and let them advise him, or even instruct him. They will have more vigour, they may not be tied so much by tradition, but they will suffer from ignorance, inexperience and from the rivalries which will grow up among themselves.

"One therefore comes to the conclusion that absolute dictatorship, and sadly also absolute democracy, are likely to fail in a developing country. What is needed is some restraining, or at least guiding, hand to prevent the father, or government, from acting unwisely or selfishly, and to save him from yielding to the undue internal pressures exercised by one or more of his sons—or what is commonly known as 'industrial lobbying', which I know about because I've done a bit of it myself. Nevertheless, he must take the final responsibility for deciding how his family wealth must be spent or invested. And that is where his wife comes in. For the purpose of my simple illustration, the wife must be like the good French *mere de famille*, who holds the family bank and knows how the joint income can best be spent, and the wise father heeds her advice.

"In the case of a developing country, the family wealth for many years to come is bound to stem from two sources; that is to say, partly from internal sources, deriving from taxes and accumulated capital in private or public hands, and partly from external sources consisting of financial and technical aid, foreign loans and foreign investors, joint or otherwise.

"The wealth must be well used, and here Professor Kaldor has given good guidance. My point is that this is best achieved if the father, or government, decides firmly that he will consult fully with and pay attention to the advice of the mother who is holding the family purse. Now who is the mother in my illustration?

"She has, it is true, a split personality. As far as the use of internal wealth is concerned, we must regard it as made up partly of skilled treasury officials, economists (including Professor

Kaldor himself who should advise on the spending of the public funds and be as independent as possible of political influence), and partly the representatives of big business concerns, both in the public and private sector. In England, to try to achieve this we have the National Economic Development Council. So far as external funds are concerned, the mother is represented by the donor or investor, government, international organizations with funds and technical personnel at their disposal, international bankers, and finally, and perhaps the most important of all, great capitalist firms, who have not only money, but technical and technological know-how at their disposal and the willingness to invest them if they are allowed a fair return."

Mr. Henniker-Heaton proceeded: "The father, or government, using freely available advice, should then be able to decide how his ten sons should be employed towards that end. And in these modern days he must, of course, consult them, the workers, through their trade unions. The manpower distribution aimed at should be somewhat similar to what it might be in a highly industrialized society today—one son in agriculture mostly run on sophisticated lines supplying 70 per cent of the country's own food and contributing also to exports; four sons in manufacturing industry; one in construction and communication; one in education and training; one in commerce and banking; one in personal services and entertaining; and one in direct government service. And actually this is very near to what it is in Israel today, because we have the extraordinary position here of a very highly developed country which is still developing at a great pace."

He declared that the developing countries which will not only attract help, but deserve it, would be those which are evaluating carefully what their best economic structure should be and working towards it—a healthy, progressive agriculture, a manufacturing industry concentrating on those products which can yield the highest rate of profit per unit of investment; incidentally avoiding export fields, such as textiles, where international competition is already too intense to yield a profit; and a training and education programme based on practical requirements.

What a developing country must not have, Mr. Henniker-Heaton urged, is a "swollen civil service, wasteful land use, expensive non-functional buildings, strangulation of initiative by high personal taxation, government by junta and discouragement of foreign investment by not giving that investment a fair deal. Nor can it for a time expect to enjoy some of the consumer service and administrative luxury which fully industrialized countries have earned by their past efforts."

On the other hand, the donor countries, the international institutes of financiers, and all those who may give aid, have a great responsibility. They must allow access to their own countries to the exports of developing countries and provide training facilities for their technicians and teachers. "But at the risk of being accused of paternalism", Mr. Henniker-Heaton observed "I must state that donors also have an obligation to insist, and a right to insist, that the aid they give is being well spent. Much of the aid given in the past has been handed out too easily, sometimes out of political motives, sometimes just out of desire to do good quickly, and a lot of it is being wasted. Political strings attached to aid are wrong, of course, but donors have a right to participate in the evaluation of the schemes in which their funds are to be expended and to inspect how they are being carried out. That's not an unfair condition at all, though it be sometimes resented by the receiving country. Some international organizations have neglected to follow up the aid they have given, and, of course, some great capitalist firms have burnt their fingers and are withdrawing, and this is a sad thing."

Mr. Henniker-Heaton did not forget the ten daughters to whom he had referred in his family illustration. He explained: "We shouldn't forget that there is a tremendous source of economic strength in the female population, if well employed and not exploited. I don't think they should be held to the 'three Ks.—kirk, kids, and kitchen'."

One discouraging experience was reported by Mr. Cyril Plant, the general secretary of Britain's Inland Revenue Staff Federation. He said: "Eighteen months ago I was invited to go to India, mainly to encourage the Indian civil servants to embrace automation, and I found that it was not only an impossible task but it was one which made me very unpopular indeed at the end of my tour. Why? Well, India has got 40 million unemployed, and I found banners at every railway station asking me to go home. They said, 'Computers in, thousands out!' and that was the message the Indians conveyed to me in very strong terms. An American computer firm had given a computer to the Indian Railways for use in their head office. Some thousands of Indian railway workers decided that they would take it in turns to lie down in the road to prevent the computer from being delivered, and they are still doing so and they have no intention of permitting the delivery of the computer, let alone it being used. If some thousands can lie down in the road doing nothing, how do you educate people? How do you get them to accept that computers will ultimately raise the standard of life?"

Mr Plant continued: "In India, with 90 per cent of the population outside the great cities, with a great strong family feeling, how can you encourage technological development on the land to drive people to the slums of Bombay where there is grinding poverty which is quite extraordinary? How do we tackle the problem in a land with 400 million population?"

He added: "These were the problems I faced, and I tried to give the message, especially to the civil servants in the Inland Revenue in India, that the drudgery will be taken out of their work, as we hope it will be taken out of the work of assessment and collection of income tax in Great Britain in the very near future, but they wouldn't have it. They would rather push quill pens than have the new miracles of the machine."

Sir Charles Wheeler endorsed Professor Kaldor's plea for improved education in developing countries. Sir Charles observed: "Not only is it desirable to provide university education, but the real need, and perhaps the most crying need, in most of the developing countries in which I have been concerned in the introduction of technical processes, is for skilled artisans, skilled mechanics. I would urge on all that the training of the skilled man is of perhaps the greatest importance in all developing countries, and I make a plea for liberalization in all manufacturing processes in our own countries towards this objective. We all have to remember in our developed countries that in certain areas—Pennsylvania in the United States, the Midlands in Britain—you have an inbuilt skill in the population which is wholly lacking in most new countries, and the only means, I believe, in getting sound technological industries into those countries is to provide training both on the spot and by transfer from manufacturing centres to countries in which plant is to be used. They are skills which ensure the maintenance of expensive plant which we, the lending partners of those countries, provide."

Mr. R. J. Kerr-Muir recalled a personal experience in Swaziland. He explained: "When it is intended to go into such a country, it is most essential that the people who are going to develop the technology should make a very intensive, on-the-spot study of the conditions there first. They should not endeavour to produce an exact replica of the plant which they have got in their home country. One thing particularly which we found out, quite different from the United Kingdom except perhaps in my own home, was that women do all the work. We found that the men had to be trained to the fact that they were going to be taking on new jobs. We found that once we had got this established, that they were fairly receptive to an intensive educational process. We found, also, that some of our young engineers had quite the wrong idea about the ability of the men to carry on the work that was required of them. The problems of nutrition in these countries is exceedingly important, and, in the costs which have to be worked out, this matter of keeping people on their feet for eight hours a day or longer is of very great importance."

Although he agreed with Professor Kaldor on the importance of introducing modern technology to developing countries, Mr. Kerr-Muir had one particular reservation. "Having seen the problem of bringing some spare parts to a country where every hammer and every nut and bolt has to be flown in over a lot of miles", he concluded, "I am not terribly excited about bringing in some of the second-hand equipment which we try to dispose of in Lancashire."

Mr. Edwin Shelley (United States), commenting on the Indian workers lying down on the railroad to prevent the installation of a computer, suggested that the people who would like to introduce computers into India might do well to think first about introducing computers into the Indian educational system, so that children who began to study from computers, "to use computers both in the course of study and to use computers in the development and solution of problems, would grow up and presumably be much more familiar with, and therefore less hostile to, the use of computers".

Lord Peddie, too, agreed on the "enormous significance" of education and training. But he had two misgivings on Professor Kaldor's paper. He explained: "I'm sorry that he did not deal sufficiently with the influence of donor countries. I have over the past 20 years had some little influence in the subject of Overseas Aid, and I've seen it at its best and at its worst. Very frequently, the wrong direction has been given to development because of mistakes that have been made in the character of the aid. I think one of the most important changes in attitude over the past years has been the realization on the part of donor countries that the provision of aid can, in the long run, bring as much benefit to the donor as to the recipient. With increasing recognition of that fact, one can secure more overseas aid."

He added: "I, too, share the view that we should not be dependent on second-hand goods. I believe that the nature, the character of the capital goods which are offered to developing countries, is of great significance."

One warning came from Mr. Dan Ram (Israel): "Important though vocational training is, it was also important not to train people who later might not be able to make their living in an industry because it had not been developed enough." He added: "So the problem is to find an integration between upgrading technological skill and the technical position of the country. May I suggest an idea which is tried here in Israel very successfully? During the period of youth from 15 to 18, it is obligatory by law, while working, to devote one day to go to a centre of training and upgrade their skills. It is an integrated system, in which young people are upgraded while producing, thus advancing the country in both fields."

Professor Kaldor emphasized in his reply that he had prepared his paper, not as "any particular expert on technology" but as an economist. And, indeed, he revealed that he had taken the opportunity to express his opinion on issues which had been subjects of prolonged controversy among economists.

He explained: "For a long time we had the view that only certain countries are suitable to be industrialized or to be manufacturers on a large scale, while other countries which are primary producers ought to remain primary producers. Perhaps we didn't quite say that it was because God had ordained it, but it was accepted that their climate and geology, their comparative advantages respecting agriculture and primary production made them suitable only for those activities, just as more fortunate countries were suited to becoming industrialized."

He declared: "I feel that it is necessary for economists from developed countries to lay that particular ghost, and make it clear as I have tried to do, that if you want to reach development in the sense of high standards of living, such as are known only in Western countries, there are no alternative roads to development except industrialization. That is to say, a country can never become a high-income country specializing on agriculture and mining alone."

Professor Kaldor continued: "How should a country industrialize? By what instruments should it specialize? The successful industrializers of this world, even though they are late-comers—and I'm inclined to include Israel soon as a successful industrializer—did it on a selective basis, and took great care that they should develop export potentials at the same time as they were developing industries which are substitutes for imports. Other countries which failed to do that, like the Latin American countries, got into a sorry state of rot, where development was seriously handicapped by grave inflationary pressures and chronic balance of payments problems, because their development increased their import requirements without adding anything to their export potential. So that was one controversial issue on which I wanted to declare my position.

"The other, more clearly related to technology, is this kind of romanticism which exists about technology in underdeveloped countries. It exists not only in my country, but in India, in Mao's China, and in all sorts of other places. It supposes that there is some wonderful secret technology to develop especially for underdeveloped countries. It has been called 'intermediate technology', and it is supposed to avoid all the unpleasantness and horrors of Western industrialization, which created large cities and urban slums and so on. It is expected to create a kind of Ruritania with a reasonably high standard living and which doesn't require as much capital per worker as Western technology."

Professor Kaldor declared: "I don't believe that there is anything in this. I don't say that it couldn't possibly be invented, but I did draw attention to the necessity for underdeveloped countries wishing to industrialize to concentrate research and development on technologies which give lesser advantages than the economies of very large-scale production. I think Western technology is slanted in favour of very large-scale output, which puts economies which have potentially a much smaller market at a serious disadvantage. But I do feel that we ought to realize that Western technological progress was not of the sort described to economic undergraduates in the various textbooks of the world. They suggest that you continuously substitute, as a result of careful saving and accumulation, capital for labour, and therefore you increase the productivity of labour at the same time as you are decreasing the productivity of capital. This kind of representation of Western technology, that it is very labour-saving, that it is very high capital-using, is in my opinion plain wrong. It is very highly labour-saving, but in terms of investment requirements per unit of production, it is no more capital-using than primitive technologies. In fact, in many cases, less so."

He acknowledged that, as an economist, he could not answer Mr. Plant's questions on the problems of India. Professor Kaldor added: "All I can do is try to give sound, rather than unsound, advice. Obviously, India is very over-populated, and this makes India's problems tremendously difficult. Surely the right advice to India is, since capital is scarce, to try to use the techniques which give the highest return of capital, and these are likely to be the most modern techniques and not the most primitive. And one ought to remind the Indians that a country has only a given employment potential outside agriculture. There are just so many workers to whom you can pay wages if they are not food producers. Anything over and above that is likely to land you in a violent inflation. Given that you have a limit to your employment potential; isn't it better to use labour more productively?"

He continued: "If you go around an Indian government office, you find hundreds of people just standing around idly outside the office of every official. They are all paid some miserably low wage and are all on the pay-roll, literally doing nothing. And this is hallowed in India as the right system because these poor chaps have got to have a job in order to live."

He observed: "This is all very fine so long as you don't take into account that all these wages paid to people who stand outside your door and do nothing, reduce the possibility of employing people outside on something else which could be of some use. The total employment potential at any given time is limited by the supply of wage-goods, and wage-goods means food. If you waste your supply of wage-goods on useless employment, then you have so much less for useful employment."

Professor Kaldor recalled the example of Japan, which at one time had similar problems to those of India. He explained: "After all, certain countries have succeeded, starting just as poor, just as miserable, and just as hopeless in their prospects from the point of view of natural resource endowment. But the Japanese have succeeded very well because they have followed the right methods; they put the right emphasis on education and the right kind of education. Instead of getting into a vicious circle, they got into the virtuous circle. By moving more and more of their surplus labour from the land, they increased the productivity of the land, they increased the surplus of the land, they increased the demand for industrial goods, and they did their industrialization in such a way, unlike India, that they developed their export potential more and more with the growth of their industry."

Referring to criticisms of his suggestion that it may suit an underdeveloped country to use second-hand machinery, Professor Kaldor acknowledged that he had introduced this idea by way of exception to his general proposition that the latest technology is almost always the one that pays best to introduce, even though it is very highly labour saving, but it has an enormous capital per worker allied to a relatively low capital expenditure per unit of potential output.

He insisted, however: "I can see that there are cases in which you can buy second-hand machinery at so much lower a price than the cost price, that for a low-wage country this is a generally capital-saving technique which it wouldn't have been if it had been bought new."

He continued: "In this case I certainly think that it has potentialities, and the fact that there is a very large international trade in second-hand machinery of all kind, second-hand vehicles, second-hand airplanes, and this and that, shows that there must be something in it, otherwise people wouldn't be buying it quite so widely. But my point about bulldozers is that the bulldozers may have the lowest capital output ratio, and therefore the cheapest technique from the capital point of view, in a Western country. They are not necessarily the cheapest technique from a capital point of view in a less-developed country where the degree of mechanical skill is less and where they may be more complex to operate and have more frequent breakdowns. So the amount of labour you require in connection with them, taking into account the repairs and maintenance, may be very much larger if an American bulldozer, or whatever it is, is used in an underdeveloped country, than in the United States."

Like so many other conference participants, Professor Kaldor emphasized that education and the development of skilled labour must go hand in hand to make these technologies successful.

But he could see no short-cuts for underdeveloped countries if they desired to emulate Western standards.

Professor Kaldor concluded: "There are no fundamentally different ways from those followed by the Anglo-Saxon countries —by the former White Dominions of the British Empire, including the United States of America which was a very early white dominion of the British Empire—and by Western European

Chapter 7

The New Colonialism

THE distinctive business institution of our generation is the international company. Interest in the emerging role of the international is at the moment still outdistancing research and organized knowledge. A good deal is being done on the organizational aspects of international management but, difficult though this may be, it is still the easier bit; there is still a bewildering blank in our thinking on political, social, and economic implications of international business.

In many ways the growth of the internationals, pushing out from one country into others, can be seen to be simply a modern version of the old mercantile system, a neo-colonialism.

These were among the many conclusions reached by Mr. Peter Parker, the chairman of Bookers Engineering and Industrial Holdings Ltd., in what turned out to be one of the most refreshing and stimulating contributions to the whole conference.

The significance of the development of this world-wide business concept was stressed by Mr. Parker's reminder that Professor H. V. Perlmutter had prophesied that "by 1988 the free world's economy will be dominated by some 300 large companies, responsible for most of the world's industrial output". Mr. Parker added: "Incidentally, for those who find these figures daunting, there is always Dr. George Steiner of the University of California, whose vision doubles; he says 600 by 1988."

But whichever academic turns out to be the more accurate, the effect will be tremendous, and it was the concern of Mr. Parker to emphasize that there is far more than cash and profit in this development.

He explained: "We are concerned with technological change and human development, and with the international company as an essential component in the combined process. Still, we would do well in considering this aspect of the scheme, to take as our premise that business is, as I think Lord Keynes put it, a bet.

"Born out of risk, the international company generates risk. For the host community there is the risk in the ambiguous blessing of new technologies bearing gifts; gain of techniques but loss of tradition; gain of momentum but loss of social stability. International companies, in their expanding activities, seem to represent unmistakably the decisive business problems of our time. Internally, they raise the problems of large-scale organizations, balancing needs for planning and controls with scope for human development. Externally, they raise criss-cross in any and often contradictory commitments of social responsibilities to both employees and the community at large, problems of balancing the claims of host and parent countries, and, indeed, of shareholders; ultimately the old question of political theory rears up; where is sovereignty to lie? It is the interplay between these internal and external demands which is now increasing attention from the business man making his living in the mixed economy of the world."

There are two main distinctions of new and critical difference between the old and new international, between, say, the Suez Company and Shell, between the East India Company and IBM.

Firstly, the imperatives of technology; and, secondly, the new management approach which is focused, rather muzzily still, on evolving ideas of the multi-national company. These two factors do not necessarily pull in the same direction.

Mr. Parker observed: "What I have called the technological imperatives might simply reinforce the mercantile system, like a powerful outboard motor attached to the old hull of economic imperialism. However, the new experimental thinking towards world-scale operations, both truly multi-national in ownership and management, wherever head office may be, I am convinced is coming. We are nowhere near there yet, but the advance in that ultimate direction has started."

He continued: "Even the glossary is confusing, and we often find that we are not quite sure what we are talking about: do we say internationals, multi-nationals, trans-nationals, or even extra-nationals? We must watch our language. For my purpose, I use the word international to describe national industrial groups operating significant investments in one or more countries. A feature of these firms would be their commitment in capital and trade spread in more than one country to the extent that the fact of geographical interdependence is the premise of their success and of their plans for expansion; global strategy and opportunities, not simply national, would be the content of their decisions. Some would say that this is the multi-national, but I save up this term to reflect the coming evolution of worldwide operations, truly multi-national in ownership and management. For these in a real sense, industrial patriotism will not be enough. Already there are examples of bits of both . . . 'inter-multis' perhaps. Beyond these crude categories I think to attempt anything more clearly cut might be misleading. We must avoid spurious orthodoxies until a lot more thinking has been done in this field."

Mr Parker visualizes some broad aspects becoming clearly discernible. And he sees them as:

(1) the influence of the international company has increased, is increasing, and is unlikely to be diminished;
(2) the international company is both an economic entity and a political entity: this dualism presents a paradox for professional management—more scope for international management, but also more involvement in political compromise, or rather, often muddle;
(3) hopefully there are harmonizing concepts of a multinational approach evolving, bred out of hard experience by hard-headed but imaginative management.

The origin of international companies may vary. They may be historical accidents or shaped by the natural pattern of their trade, as in the commodities of rubber, copper, oil, and sugar. Or the pressure may come from market forces: either the expansion of success at the home base beyond the point where a bigger market share is feasible or legal; or through the resolve not to lose an export market where tariffs are being raised.

Growth under the pressure of the market leads through familiar phases of evolution, with a rhythm, Mr. Parker added whimsically, "we might all recognize as that demanding dance routine, the fox-trot—slow, slow, quick, quick, slow.

"Phase one, slow, a trading relationship of exports and imports, with possible portfolio investments overseas.

"Phase two, still slow, transferring technical know-how, perhaps under licence agreements, for manufacture abroad, and contracting or joint ventures with minority interests; and at this stage still no managerial commitment, only the flow of goods and cash and know-how.

"Phase three, the management pace quickens with the setting up of manufacturing facilities and sales organization.

"Phase four, with management right on its toes, the setting up of a full range of management components and enterprise, all the functions but operating within the international strategy set by headquarters.

"Phase five, cannot be hurried, slowly does it, the crucial adjustment to the environment."

But whatever the origins, the fact is that a modern, expanding enterprise, if it is to be true to its nature, if it is to be effective and profitable, must think and act globally. Isolationism, nationalism, is proving unworkable in this generation of business. The British have recognized the force of this logic in its relation to the Common Market. And there is little proof of the diseconomies of scale among the great internationals, rather the opposite. Take the case of modern enterprise not only in frozen foods but in computers, in cars, in aluminium, in man-made fibres, in aircraft. How can the cost of selling, research, production planning, services, people, be endured if they are not spread over many markets? The economies of scale are not so much a matter of size of plant. They are related more to the strategy of human resources of management and to the fundamental needs of research and development, seen in the round as meaning both production and marketing, with a special emphasis on marketing. This is relevant to some producer goods industries where technical research and development forces the pace, but particularly so to the producers of consumer goods who have often blazed the international trail for business in modern times, because it is in their field that economies of scale in research and development really pay off.

Mr. Parker added: "The trend seems obvious enough; the needs to be served by business enterprise may be fulfilled more naturally in global rather than in nationalistic terms. Any company solely based on its own country is likely to be driven to the wall by that scale of competition. A tariff wall, perhaps, but in terms of international competition it is dead."

It is possible that 200 of the 300 dominating firms mentioned by Professor Perlmutter will be American. Already the impact of American investment, finding that Europe offers the best expansion potential, has caused a shockwave of concern on the Continent. Already the rise in the USA share of international companies is overwhelming. Before the war foreign investments of companies engaged in international business was 15,000 million dollars. Now it is 100,000 million dollars, and is still rising. The total book value of the foreign investments of USA companies in overseas affiliates amounts to about 60 per cent of the total.

According to recent US Department of Commerce figures, the total book value of USA foreign investment in Europe was matching that of their stake in Canada in 1966. Mr. Parker commented: "To some European business men Canada appears increasingly to be a North American attic. And there is no relish here in seeing Europe becoming an American outhouse."

American companies, for instance, already control 50 per cent of the supply of semi-conductors and 80 per cent of the computers in Europe. In France, American companies control 40 per cent of the petroleum industry; 65 per cent of the agricultural machinery; 65 per cent of telecommunication equipment; and 45 per cent of synthetic rubber production.

Mr. Parker recalled that in his book *The American Challenge*, M. Jean Jacques Servan Schreiber premised that the third most powerful force in the world within 15 years, after the USA and the USSR, may be American industry in Europe, and he, M. Servan Schreiber, called for a European counter-offensive, in which he appeared not to have all that confidence.

Mr. Parker continued: "The paradox is that the internationalism of enterprise can so exacerbate nationalism. Yet what European business and national sensitivity is up against is not a cunning, co-ordinated attack by an exploiting power. We are facing an onrush of entrepreneurial exuberance and formidable management capability, powered by accelerating technology with advanced systems of information. The Europeans now have the taste of what Africa, the Middle East, Asia, and South America had to swallow when we expanded into them in the heyday of the nineteenth and early twentieth centuries. Only this time the expansion is not based on trade but technology, and it is not trade following the flag, but the flag following technology."

Nevertheless, Mr. Parker points out that it would be a mistake to distort the significance of the role of international management by too obsessive a concentration on the present predominance of US based groups. The problems which are being thrown up by the development of international companies are not simply met in the United States. From the "Fortune" list of the 100 manufacturing and mining companies with sales of more than one billion dollars, some 50 had at least one-quarter of either assets or sales overseas, and 23 of the 50 are European. Also, the growing element of European investment in the big US based groups is a factor often left out of the calculations.

What we must keep centrally in mind are the global issues of relationships between enterprise and the sovereign state, between the international company and the local community in which it operates. Of course, tensions on this front are more acutely registered between the developed and less-developed economics.

The American company in Scotland, or the British company in Germany, has less need to justify itself as private enterprise, less difficulty in "nationalizing" its local management and politics.

Mr. Parker concedes: "There is always the temptation to exaggerate the effect of the internationals. Nowadays we seem to be at a stage at which the international enjoys almost romantic respect. It has all the thrill of the 'big bang' theory that will explain everything—how we can make the most of technology and the least of international divisions; how with increasing industrialization we can make continual change both welcome and profitable; how we can 'order the unknown'. The truth is that in terms of their effect on the overall economic pattern of world trade, any aggregates prove to be relatively tiny. Even culturally the case of the 'ugly manager' may be taken too far. (Maurice Zinkin in a recent article deflatingly wielded a rapier of ridicule against the blown-up image of internationals; 'no western country is going to lose its independence because its cereal is made by a foreigner'). Yet, that said, what matters surely are the marginal pressures, for instance:

"On the jugular vein of sterling when internationals hedge against devaluation of that currency:

"Or when General Motors decides not to export from its Australian subsidiary to Japan, but to leave the Japanese market to its Californian plants. Australia had no other Australian produced car to send to the Japanese market and Japanese cars were flooding into Australia;

"Or, to take a less dramatic instance, the sort of continuum of social confrontation of the copper companies in Central Africa;

"Or, in even more day-to-day terms, take the often feudal contrast in a developing country between the living standards of international management and those of the community in which it operates.

"Ultimately it is these marginal issues which prove to be the combustible emotional charges in relationships, and these relationships, so coolly described in the jargon as environmental factors, are critical to the success of the international. Without an understanding of them, or perhaps equally important, an attempt to understand them, without what Adam Smith described with his canny accuracy as 'sympathy', most professional international management can come a cosmic cropper. This, then, leads us to another paradox of the international business world: more professional, more technologically based management—and yet more political involvement.

"The standard reflexes of any manager are to keep out of politics, and by nature management is as apolitical as an amoeba. The crux of management is the setting of objectives, with precise definitions of delegation for devolution, of decision-making to the lowest levels and for individual accountability. These are familiar principles for any scale of management, but their relevance to international organizations sticks out all the more obviously. The extent of central corporate control will vary with the type of international organization; usually one of three basic patterns—functional, regional, or product-orientated, and with the type of industry. On the one hand, there is the business based on a unified technology, working a complicated international system of supply areas for goods and services and know-how. And on the other hand, there is the conglomerate, dealing in many products. But the decisive point is that with the breath-taking improvement in communications, central corporate control becomes even more feasible and firm and tempting."

Mr. Parker commented: "The new world may or may not be braver, but certainly it will be smaller. And the temptation to centralize more and more corporate decision-making is vastly strengthened by this."

Old fashioned management of international operations in the past, he continued, sometimes placed itself, and to be fair was often placed, on a pedestal of paternalism. Is there a danger now that paternalism is to be programmed on a computer? From all the IBM is doing, of course, it is obvious that that is the one thing they are planning the computer will not do. But the general point remains. The imperatives of technology can surely reinforce inflexibility in management, and even create a new version of paternalism.

Mr. Parker insisted that it is in the stresses and strains between corporate strategies for large-scale organization and the environmental factor of operations—the social and political context of making profits and losses in somebody else's country—that the international has to establish its role. And ultimately it is the achievement of local acceptance which is decisive.

He reported the experience of one of his own companies in Africa.

Mr. Parker recalled: "In front of the fabled black rock of Mulungushi in April last year, the President of Zambia, Kenneth Kaunda, announced that his government was taking over the capital of twenty-five European companies in the country, including our Booker department stores and shops, with capital employed of more than £2 million. This for Bookers was a shock to our system, but not a complete surprise. In fact the moderate mood was typified by the reaction of our African director, who, when he heard the news of the take-over, actually said, 'Why didn't they ask me, and I would have stopped it?' Anyway, we have responded as positively and co-operatively as we can in the formation of the new Consumer Buying Corporation in which ownership will be vested 51 per cent in the Industrial Development Corporation of Zambia, and 49 per cent in Bookers (Zambia). We now have a 10 year managing agreement signed to manage this new extended Corporation, and we have every reason to believe a fresh and profitable chapter has opened.

"This is not a desperate story compared to those who have been through tormenting times, say, in Indonesia and Congo. It was, as one left-wing commentator puts it 'much the sort of thing an undogmatic but dominant European Social-Democratic government might do if it found itself in power in an African country'. Yet it has a bearing on the attitude we have in Bookers to our international risk-taking."

He continued: "Without being too pompous, I must try to give you a personal outline of our philosophy as an international:

"1. We believe that the continuing optimum profitability of our enterprises not only fulfils our responsibilities to our shareholders and employees, but is the best measure of our effectiveness in the service of the community in which we operate.

"2. We try to offer opportunities—and I believe we have genuinely succeeded—to members of the local community to make careers and share in policy making at all levels of our enterprise; we recognize a responsibility in developing educational and training facilities, both inside and outside our industries. (We have African and Guyanese directors, but none on the parent Board.)

"3. We are keen to provide attractive opportunities for local investors in our business, and at the same time, therefore, we encourage some form of market for the shares. (We have, for example, formed Guyana Industrial and Commercial Investments Ltd. and have direct local shareholdings in three of our operations there.) But you have to make profits to make progress on this front—for instance, much as we would like to go public in Trinidad, we are not profitable enough there yet.

"4. We try to work with the government of the day: and in doing so we accept the concepts of the mixed economy as a basis not only of coexistence, but constructive and inevitable partnership."

Mr. Parker added: "Statements of this sort, at conferences of this sort, tend to sound like the bland leading the bland. I do not pretend that this fourfold approach is written on tablets of stone—nor that it is definitive, and I am conscious too that a conglomerate can be much more easily flexible in its approach than a more technologically based industry. But with that said, it is still a help. It helps a company stick the pace, especially when such a moment of home-and-overseas truth comes as it has for us in Zambia. I am sure that without it we should not have been entrusted with the new scope of operations we now have in that country. Also, it is an example perhaps of the way in which an international business, exposed as it is in developing countries to a very high degree of visibility, is driven to declare its position more explicitly than the home-based who can take it much more for granted. The international must prove capable of economic osmosis, of accepting the inter-penetration of business and its environment. Its capacity for local demands and for identifying itself with the destiny of its host-community is the key, not only to its success, but sometimes its survival. The balance of this priority with that of developing the new and necessary breed of international managers which we are hearing so much about, calls for nice judgement. This new order of manager, represented by the ideal of a multi-national team, expert and mobile, the true professional, can look to the eyes of an African awfully like a technocratic eunuch; a watchful guardian over the assets, but incapable of commitment."

He continued: "International management cannot explain itself to its host country, advanced or less advanced, in managerial terms alone. Yes, we are an attractive package, which mobilizes capital and advanced technology and wider marketing skills, all tied up with a community relations policy. Yes, we are an inevitable partner in the general development effort, in association with Aid programmes. Yes, even the purpose of our profits is becoming better understood in the process of mutual acceptability; without a sound return on the capital to be invested, very often with special risk factors involved, how is big business to justify the capital it needs to make its contribution towards development expenditure? Yes, all this is a lot, but not enough. Without 'sympathy' it will not work. The international manager does well to remember the Confucian remark, 'Why do you hate me so much, I was never kind to you?' The international cannot appear to a host country as a sort of self-contained black box with a happy knack of survival whether there is an economic take-off or a crash."

Mr Parker declared: "This raises the final and delicate issue of commitment: ownership. The hard-liner of managerial logic, say, General Motors, may well hold no compromise on 100 per cent control and ownership. But two questions ask themselves; the first: If it is wrong, say, for General Motors to have half of every car market in the world, who will say so? Secondly and bluntly:

Are people likely to stomach the outside ownership of their key assets for longer than absolutely necessary? There is no doubt that the time has come for some international review of the tangled legal field opened up by the first question. The human reality of the second question challenges the energy, flexibility, and the invention of an international."

Mr. Parker concluded: "International management has a vested interest in surmounting barriers, of thought as well as frontiers, and that cannot be a bad thing."

First comment on Mr. Parker's paper came with great authority from Mr. Edwin Ronald Nixon, managing director of IBM United Kingdom Ltd., and chairman of IBM World Laboratories Ltd.

Mr. Parker, he said, had been right to talk about the imperatives of technology as a crucial distinction between the international companies of today compared with those of the last century. Mr. Nixon added: "Within IBM, from my point of view, this difference has been emphasized in the last 15 years. In the middle 1950's, each IBM factory in Europe, by and large, produced the total product range of the company. In 1957 we introduced a policy of what we call 'product by plant', and this meant that each factory specialized in the production of a small segment of the total product range, and then exported to all other countries in Europe, and, indeed to the rest of the world. We have passed through a number of different technologies. We have progressed from valves, through transistors even to MLT. The cost of seeing to the setting up of a production line for a component in this latest technology is such that we can only afford to have one such line in Europe. This means that a typical IBM plant in Europe is now essentially doing two things; firstly, it is making a series of components and sub-assemblies, and these are shipped to other IBM factories throughout Europe, and, secondly, it is responsible for the final assembly of a certain small number of completed machines.

"For example," he continued, "in our factory in Greenock, in Scotland, we make the computer memory units of the system

360 which we send to France and Germany, and these memory units then get incorporated in the final assembly of those computers, which get shipped all over the world. We also make certain peripheral units at Greenock, and a computer ourselves which we send all over the world. In essence, the new technologies have meant that we have been forced to ignore the artificial tariff being set up by individual countries, so that when these tariff walls break down, as indeed they will, the only impact on us is that we become more profitable.

"I want to comment on Peter Parker's reference to 'global vision'," Mr. Nixon proceeded. "It is very easy to say, but I don't think that any businessman today can live without this global vision. Products and services have to be designed to meet the requirements of a world market. Modern travel, modern communications are making the world a smaller and more homogeneous place. The Volkswagen and the Mini-Minor are seen all over the world. This contraction of the world is partly responsible for the forced mergers which are now taking place in many industries, but especially the ones involving the new, higher, and very costly new techniques and technologies. It is also touching on a point made by Mr. Parker about risk-taking. Certainly, business is a risk. It is part of the job, and part of the joy, but the costs of mistakes all the time are getting higher.

"Let's take the aircraft industry for example. There is a limited world market for aircraft with a certain range and certain capability, and the specifications of the customer, the airlines, are becoming more and more stringent, so that, in fact, only two or three companies in the world can meet their requirements. But what happens when only one or two companies get orders? The fact is that the third is going to go out of business or merge, or certainly lose its previous format. When the British Aircraft Corporation, for example, put the initial investment into the Vickers Viscount, the cost was about £25 million. The investment costs for the Concorde have soared from an initial estimate of perhaps less than £100 million to something over £400 million, and that may not even be the end. Even allowing for some duplication because two countries are involved, it is still a substantial degree more than the original £25 million for the Vickers Viscount, which, incidentally was an outstanding success. The investment cost of the BAC 311, the British competitor to the European Airbus, is probably in excess of £400 million. The break-even point in the sales of the aircraft implies a major improvement in the world market share of BAC. It takes not only managerial skill and technological skill to take those decisions—but it also takes guts."

What Mr. Parker had called sovereignty, Mr. Nixon preferred to describe as "conflicts of interest". He explained: "I am responsible for all IBM operations in the United Kingdom. What happens when, in the interests of the IBM Corporation as a whole, a certain decision shall be taken which conflicts with the interests of the British economy? My point of view, which has been forged as a result of experiences in certain practical matters, leaves me in absolutely no doubt that the interests of the Corporation must be submerged in the interests of the country. That is my position today. As the march of the multi-nationals progresses, with national boundaries becoming less rigid and less meaningful, although recognizing that there is still some way to go, and with the development of ambassadors and courtiers working under new international laws and environments, I believe that I could change my mind. Today, however, it is clear. The host country's interests are paramount, and indeed this causes conflicts within international companies."

And Mr. Nixon gave an example. "Some years ago," he said, "IBM decided that if we were to be successful in South America we had to have a factory there. What should it make? The Corporation decided to make sorters, which are devices which sort punch cards, a fairly simple electro-mechanical device which we happened to be making in Scotland. We stopped making the sorter in Scotland and it is now made in South America. Instead, in Scotland, we are making a product which is demanding a higher level of technology than that sorter. This meant an upgrading of the skills of the work-force in Scotland, and incidentally it also increased our export values. That is fine. What is the next step? South America is an expanding market and we need more manufacturing there, so we decide to make some more peripheral equipment which again, coincidentally, we happen to be making in Scotland. The volume, however, is such that for a period of years we can have two production lines, and this will give us time to readjust in Scotland to prepare for newer products with all the associated retraining of skills which are required. Some of our people in Scotland in the last 10 or 15 years have been trained from shipbuilders to mechanical engineers to electro-mechanical engineers and now to electronic engineers."

Mr. Nixon continued: "While these changes have been taking place, my sister companies in IBM France and IBM Germany, for example, have also been upgrading their skills, and now comes the crunch; a new technology comes along and only one production line will be set up, and clearly there is an obvious export potential. Where does it go? As I said earlier, the risks and the stakes are getting higher, so that suddenly I am locked in battle with my colleagues in France and Germany to get the Corporation to make a decision to make a particular new technology in the United Kingdom. The decision is by no means simple. It does not become merely a manufacturing and sales decision, but it becomes a political decision in the widest possible sense. I agree with Peter Parker that the businessman's natural instincts are to shy away from politics and government, but as business becomes more complex, the interests intermingle and businessmen are forced to become involved. The interests impinge on increasingly sensitive areas."

On the problems of control and ownership of international companies, Mr. Nixon observed: "As communications and computers develop, and in spite of the wider activities of business, it becomes more possible for world-wide operations to be managed from a central source, usually sited in one country, although there are examples where that is not so. Central policy making body is responsible for developing the strategy for a company, issuing guide-lines for subsidiary companies operating elsewhere in the world, but leaving the tactical operations to the management of the subsidiaries. This raises issues about the authority and responsibility of those subsidiaries, and the personnel problems involved.

"But I want to focus on the relationship of the government of the host country and the operating subsidiary, and in particular on the subject of where decisions are taken and where control actually is. Independent decisions taken in Detroit to start Ford and General Motors activities in Belgium, for example, a few years ago clearly had a tremendous impact on the Belgian economy. It is necessary, therefore, for the headquarters of an international company to be extremely sensitive about taking decisions centrally which have an impact on host countries where they operate. How can this sensitivity be assuaged? Frankly, I don't think it can, but there is a belief that local ownership or local participation can materially assist, and here I would like to praise the policy of a number of British-based international companies, who, as a point of principle, insist on local participation in the ownership of the local company. Peter Parker's company is one, RTZ is another, and Dunlop a third. This viewpoint is diametrically opposed to major US-based international companies such as Ford, General Motors, and possibly my own company, IBM. Who is right? Time will tell; but I believe that the underlying trend is for the British attitude to be in the ascendant."

Mr. Nixon acknowledged that about nomenclature—international companies, multi-national companies, trans-national companies—it was very difficult to be precise, but there is a broad spectrum from the company which operates from a home base with occasional operations in one or more countries through a worldwide operating company to a worldwide company operating without regard to national boundaries or the nationalities of personnel involved.

He added: "We talk about the multi-national company where essentially the best men get the job, regardless of their nationalities. The companies still have a total accountability to one home country. It is a fact that some of the profits of some of the largest companies in the world are greater than the gross national product of some of the smaller countries in the world. It can be seen, therefore, that board decisions taken by these international companies can have more impact than decisions taken by some of the governments of smaller countries."

Mr. Nixon concluded: "I see the future, therefore, as one in which international companies will become more alert, not only to their business interests but to their responsibilities to mankind as a whole."

Lord Peddie, who is a deputy-chairman of Britain's National Board for Prices and Incomes, recalled that his board had been involved on two or three occasions with problems arising from the pricing policy of international companies. He invited Mr. Parker to express a view on the advisability and possibility of agreement between national governments on the subject of a pricing policy for international companies.

The discussion was broadened by Professor Ben Roberts, of the London School of Economics, who anticipates that the "extremely rapid growth of the multi-national corporation will be a dominant form of economic organization by the end of this century". This development, he added; "is going to have profound repurcussions for collective bargaining and for relations with the trade unions because the workpeople in these companies must be extremely concerned by the kind of decisions which have to be taken by the international companies."

Professor Roberts continued: "We have already seen this in the way in which the trade unions in the United States have responded to the development of the motor-car manufacturing companies of that country in terms of the expansion of their overseas plants. Last year the International Metal Workers Federation convened an extremely important conference in Italy to try to work out a common approach cutting across the boundaries of nations to deal with these problems. I think there are other trade unions seeking similar arrangements at the present time. This gives rise, however, to conflicts within, not only between the unions and the companies concerned, but conflicts of interest between the unions of different countries. It opens up a whole range of new problems that the unions will have to concern themselves with in the future. I think they have only just begun to discern these problems. It would mean developments in terms of union organization; it would mean that unions would have to develop new departments with new types of expertise and new flows of information. And one of the problems in this field, so far, is that we have hardly any information about comparative operating costs. I think the companies themselves know this, but we are likely eventually to get a public demand for more information across frontiers in terms of comparative costs and so on. This will present the management of multi-national companies with a very considerable problem."

It was unlikely, he thought, that the types of decisions mentioned by Mr. Nixon, made unilaterally by the managements of corporations, would be permitted eventually because of the growing demand for a sharing of information. It was possible, too, that a redefinition of international codes of conduct would be required. There is an International Labour Code, but it was devised in the past in the form of setting up minimum standards of a particular kind.

Shipping is already on the way to becoming an international business, suggested Mr. Douglas Tennant, the leader of Britain's mercantile marine and airline officers. He explained that the cost of a modern, sophisticated container carrier, carrying say 1200 containers at a speed of 21 knots, is from £4.5 million to £6 million. He continued: "It is estimated that one container ship can carry the same profit as four conventional ships, and that guess is probably more or less right. Such being the case, more and more countries are desirous of having a share, for balance of payments and other reasons, in the carriage of their traffic, imports, and exports. And we are now seeing the development of the International Container Consortium. Now the United States, Sweden, Holland, the United Kingdom, and France have got together and are planning identical equipment. They are sharing the cost of the equipment and, one assumes, that the share of profits from this development will be related to the amount of capital invested. They are also getting together to share out the traffic between them. We are seafarers, acting through the International Transport Workers Federation, and we have got together and agreed what the basic conditions of employment should be. We will not be pawns in the competitive game, and we want a stake in getting some advantages from this new development."

Mr. Tennant disclosed that the International Transport Workers Federation is holding a conference this year to deal with "vocational training in this sophisticated shipping age", in which the change is "probably the biggest since the change from sail to steam". There is also an attempt to get a convention to "deal with the social implications arising from the application of automation to ships". Should there not be a greater attempt, asked Mr. Tennant, to use the International Labour Organization to solve some of the "chronic social problems" likely to arise from these developments?

Sir Charles Wheeler, a director of important British engineering concerns with overseas interests, acknowledged that one of the principal instruments of technological development will be the international company. He said: "Personally I have had a great deal to do with the development and creation of companies overseas from large manufacturing companies based in Britain, and I have always been rather critical of the attitude of the big American corporations towards local participation. I have always held the view that local participation is very important, whether it be a large participation by locals or even a majority. In one case, I voluntarily gave a majority situation to the local country. But one great difficulty facing technological industries is the problem of the pricing of the exchange of components, and particularly of know-how."

One completely new phenomenon is happening in the world today, concurred Mr. John Hargreaves, a British executive of IBM, and that is the growth of these large corporations. This means the growth of a unit "which has not only business, but political, social, economic connotations as well, and because this has grown up into what is becoming an industrial state, the government of that state finds itself in potential conflict with the government of national countries. The difficulty is that they don't play by the same rules, because one is international, non-geographical, and the other is physical, national, and geographical."

Mr. Hargreaves continued: "Because of this, and because it is a fairly new phenomenon, much of the dialogue between the large corporations and the national countries has until now been on a complaint and pressure basis. But we haven't really ever sat down to define interests, to talk about working out schemes whereby we can live together. I am sure this present basis of dialogue is going to change, and we shall find ways in which some of the present responsibility of national governments may be handed over to the large international corporations. There are two areas where I can see these two powers working together: first of all in our home countries, in the field of education. In my own company, out of some 10,000 employees, more than 2000 are graduates or equivalent professionals. This is an enormous ratio of expert, skilled people, and I'm sure that we shall find ways, practical ways, in which some of this expertise can be made available within our countries in schools, training and universities, and so on. The second area of co-operation must be in the developing countries, and here there are whole fields of environment and science where this same expertise can be made available. One thinks of water resource control, scientific farming, crop analysis, pollution, education, and medicine. All these are fields where the expertise at present available can be made over in practical forms to the host countries. This is the kind of study which might well be undertaken in detail by working groups representing interested parties."

Lord Douglass of Cleveland recalled that the international aspect of industrialization had been carefully considered by the International Secretariat of the Trade Unions for many years. He added: "The American trade unions brought the problem to us, I think 15 years ago. They suggested that what would happen throughout the world would be what happened in America; it was the flow of investment from strongly unionized areas to the weakly unionized areas, where wages were low."

He recognized, however, that the wages cost is not so important as it was 20 years ago, because the capital cost of equipment has now risen to such heights that the labour costs can be negligible.

Lord Douglass added: "I myself took out a simple arithmetical calculation on the movement of a bank rate at 3 per cent. If you have a loan of £100 million it is simple arithmetic to say that 1 per cent would cost the borrower £1 million. On a loan, therefore, of £100 million, with a bank rate charge of 3 per cent, there can be a difference in two countries where the bank rate varies by 3 per cent of £3 million. Now these are fairly cold figures looked at as statistics. But if you convert them into wages, £3 million become the wages of 3000 men at £20 a week."

He asked: "If investment will not flow to the countries where wages are low, will it then flow to the countries where the bank rate is low? What effect will the bank rate have on an investment? What effect would the investment have on the bank rate?"

He concluded with yet another question. Lord Douglass observed: "We have seen the effect of a strike in one country on another country where you have international industrialization, and you can have a strike in one country which can stop work in another or a number of other countries. I don't want to go into the reasons for strikes because you can have them generated for the purpose of causing difficulties in other countries. Will the investment tend to move to countries which have the fewest strikes?"

Intriguing notions of some sort of "secret society" of the world's great international companies were inspired by some of the calculatingly guileless questions put by Mr. Cyril Plant, the general secretary of the British Inland Revenue Staff Federation. He confessed that his eyes had been opened by Mr. Peter Parker and Mr. Edwin Nixon, because he had not realized the ramifications of multi-national companies. He had always understood multi-national companies to be concerned with such primary products as rubber, oil, and sugar.

He continued: "But rubber, oil, and sugar are in countries where there are some upheavals from time to time—Cuba is a case in point—and nationalization could be feared by multinational companies. But you just can't move a sugar plantation so easily. But if there is a threat of nationalization in, say, Scotland, these companies could move their factories elsewhere, creating unemployment but at least taking their know-how with them. So, therefore, when you get the new technologies in advanced countries, new problems are raised. How prepared are the multinational companies? Have these companies got a club to which they go to exchange ideas on such things as how far profit is retained in the host country? Is there a sort of understanding between them? Is there a common labour problem? Is there a cross-fertilization of ideas and men, and does one company try to take a little bit of advantage over the others or do they march hand-in-hand? Have they got an understanding on what would be a consortium of action in relation to nationalization?"

Clearly, these questions were not a reflection of Mr. Plant's own experience, for he was invited some years ago to Guyana by the Guyanese TUC to advise on a dispute between local workers and Mr. Parker's company. Mr. Plant explained: "This was when there was a Commission into the problems of the sugar industry and the local trade union members in Guyana thought that Mr. Parker's company ought to be making much more profit than they said, and could we look at their accounts and be quite certain that they were speaking the truth. Well, I'm bound to say that Mr. Parker's company was terribly helpful to us, and we, after looking at their side of the story, understood the position very much more clearly and were able to give some sound advice to the Guyanese TUC and the disagreement was settled reasonably amicably."

Mr. Plant concluded: "I mention that because the international companies, to my mind, are not afraid of speaking quite freely on some of the difficulties they come up against."

Mr. Yosef Ami (Israel) declared: "The strongest impact on our industrialization was made by Israelis trained overseas and sent back home by foreign investors and used here as their local managers. These were the most successful people to translate foreign industrial experience, know-how, and managerial practices and adapt them to the Israeli reality. I have in mind a few cases of outstanding successes, like Monsanto and others. These are the companies which know how to use local talent, take them abroad, train them, bring them over here, shorten the period of running in, limit the risk of trial and error, and use Israelis as a link between them and this country. But there is another aspect—the local investment of profits. As an Israeli I would like us to make the greatest profit from economic activities, especially industrial activities, in our country. There are only a few of these companies who are clever enough and successful enough in re-investing here, especially in local research. Not all companies realize that their image is assessed by the way they use their profits. Nor do all companies know how to develop a local infrastructure by encouraging local technical services and pressing these technical services up to international standards, developing small industries here to supply them with components and small parts."

Mr. Harry Briggs, an executive of the Unilever group, recognized that only by employing nationals of a host country and by training them, sometimes for very senior jobs, could the aspirations of that country be satisfied.

"In the company that I work for", Mr. Briggs continued, "which employs about a quarter-of-a-million people around the world, we have something like 15,000 managers, and in the Continent of Europe we have at any time, something like 70 per cent of the continental managers working outside their own country. The object of that is to try to bring a truly international aspect to their job and, indeed, give them international experience. As far as the developing countries are concerned, excluding West Africa where we have a pretty big operation, while the total management strength grew from 1000 to 1500, the percentage of expatriate staff fell from 23 per cent to 13 per cent. In West Africa, where we have something like 1000 managers, 49 per cent of those are now Africans."

In his reply to this far-ranging discussion, Mr. Peter Parker emphasized the need for a great deal more research into many aspects of the development of international companies. He added: "What material there is, is very thin, beaten thinner by repetition."

One of the themes which kept being repeated throughout the discussion was the inadequacy of management simply as management. Mr. Parker commented: "We talk of management as a profession. That somehow or other produces a man who is as skeletal to me as the idea of the economic man, who is concentrating simply on his functions and on his profits. It is a difficult concept if you are really dealing on the scale of the problem which we are thinking of in the future. We have to consider the range of social and political implications behind all major management decisions. We must see that our management education programmes are not simply in the technical disciplines of management."

Mr. Parker declared: "We shall not fail for lack of technology. We might well fail for lack of awareness and sensitiveness, and I feel that each individual company struggling to get rapport with its host community must never assume that it has got the answer, and must continually be feeling its way towards new approaches and new ideas."

He continued: "Those of us who operate in developing countries are deeply conscious that we have to help to define a philosophy towards our profits, and I think that this is the second area for research. Very often industry is a bit shy about talking of its overall problems of profit, and to the national it can often seem inconceivable that the foreign investor is not doing just as well, if not better, than the local manufacturer that he knows about. The silence of international companies on sensitive issues, and profits especially, seems very often to confirm this suspicion. And this unease, of course, deepens very often when national debates on balance of payments take place—such as those which are taking place in my own country and, indeed, in the United States. Corporation after corporation has presented evidence to the effect that the outward flow of profits and royalties exceeds the outward flow of capital to the developing countries.

"So we have a paradox, and I would have said that the challenge we must put to any deeper thinking is: Has an international company got a definable philosophy? Towards profits? What does it consider it can pay which would be mutually acceptable? Is it the best basis for the success of an international company in a host country? What is a reasonable profit? And with considerations of oversimplification apart, I feel that we have got to have something to say on this, whether we say it is 10 per cent after tax on capital employed, justified on the grounds of normal business, where the need for the ordinary shareholder is at least 5 per cent, with a need to put at least 5 per cent to reserves."

Mr. Parker added: "But the paradox is that a company operating in an area where the politics and social priorities are combustible is going to call for a greater return on its risk capital. Lord Douglass raised these issues and asked what we could do about considering whether risk capital is going towards the countries that have a cheap labour rate, a low bank rate, and a low strike rate. Well, it is interesting that if one considers the record at the moment, one would see that private investment is not making that substantial an increase in its flow of investment to those developing countries, where some of these attractions may seem to be there at first blush. I think that since 1950, private overseas investment has played a much smaller role in resources to poor countries than official aid.

"But these attractions are, I believe, not really the decisive inducements to a company that is thinking long. If you are going to be operating internationally, you are thinking of a time scale of 15–20 years, and I don't think the smaller questions—though they have an immediate effect—are likely to be decisive on the long-range strategy of success."

Mr. Parker expects that at some time or another a code of practice for international enterprise will be produced, and it will cope with such problems as social policy, profits and the remittance of capital, and standards in joint ventures. But we are at present in an "extremely transitional stage". He recalled that the United Nations held the first session of its Trade and Investment

Conference in 1964, but there was no club to settle these issues. And the International Chamber of Commerce is trying desperately to produce some kind of code of practice which could be acceptable.

He continued: "Also we see how in country after country, like Canada and the United States, there are understandings coming on anti-trust action. We see a great deal of new thinking in Europe. But I think this would be only a partial development. We are dealing—when you think of the changes in the world's map in the last 10 or 15 years—with countries that are establishing themselves with all the assertion that countries have in their early days. And yet these countries have emerged, say in Africa, often with no logical frontier and without the logic of an economy. We may well have to wait for a lot of these small countries to coalesce into regions, and within those regions to produce some progress."

Mr. Parker declared: "I believe the international company is going to become the focus of much new thinking. Of course, in doing this, we must be asking ourselves all the time, how do we justify ourselves to our host country? All right, we are efficient; we are the attractive package that can offer management as well as capital, and can offer management in educational training; we can offer a broader base for savings and domestic development; we can offer, perhaps, some effects of entrepreneurial flair.

"We can offer all those things, but can we offer with them, too, a clear philosophy of approach? That is where we start to fumble. Efficiency to what purpose and at what price is the key question, whether we are in an advanced economy or an advancing economy. Industrialization does not seem to me, advancing economy does not seem to me, to be coming up with tidy answers. We are left with the feeling that there are a whole set of paradoxes which we haven't got answers to."

Mr. Parker concluded: "On the one hand, we can see more skills, yet more differentials; we can sense more equality yet more élitism; we can see more fellowship, we can see more loneliness; we can see more freedom, we can see more servitude; we can see more security, and yet more inter-dependence; we can see more civilization, and yet more barbarism—and these paradoxes are at the heart of industrialization. To think that we can find an answer to all these paradoxes is totally unrealistic."

But he ended on this note: "We can tilt the movement one way or another and I think that a conference like this does help us. If we can say, after even a broad discussion as incomplete as this, that if we are not where we should be, at least we are not exactly where we were when we started the conference."

Chapter 8

Fairer Shares in National Wealth

THE time has come to consider how workers should have a greater share in the ownership of capital. There has been too much concentration in the past on how to increase wealth and not enough attention on how to share it more equitably. The neglect of the problem of distribution of wealth has gone too far.

If the Western democracies believe that ownership should imply effective control, and if we are right in thinking that property ownership will become relatively more important in the next few decades, there is need for a new kind of institution in the capital market which will promote the ownership of assets by employees, and be strong enough, in terms of the control of investible funds and company voting rights, to ensure that they are administered in the interests of employees.

Automation may well turn out to be labour saving in the strict sense of the economist as compared with popular usage, which means that it may be so much more profitable to use capital in place of labour that the input of capital into production will actually rise. The share of labour in a rising national income will fall, and unless employees acquire claims on the increased property income, there is clearly a grave risk of alienation and aggravated conflict between the propertied and employed classes in advanced capitalist societies.

This was the theme of a remarkably detached and analytical paper on "Incomes policy and sharing in capital gains", presented to the conference by Professor J. Rodney Crossley, who is a professor of industrial relations at the University of Leeds.

In some ways he was breaking new ground. He did not quite live up to the light-hearted introduction of the session chairman, Mr. Bertram A. Powers, president of the New York Typographical Union, who quipped that an alternative title might be "How to increase your share in the swag".

Professor Crossley, a former adviser to the British Ministry of Labour and the Prices and Incomes Board, had no facile solutions to the problems he posed.

He was content to provoke further examination, and willing to be controversial. But at the conclusion of the conference there was general agreement with Mr. Theodore Kheel, the leader of the American delegation, who declared that the consequences of Professor Crossley's analysis were so far reaching that they might well turn out to be one of the most important contributions in recent years in the study of future technological developments.

Some of the shrewdest participants at the conference acknowledged that Professor Crossley had been a pioneer in thought, though there were others who confessed that they found some difficulty in following some of his more abstruse arguments. But the gist of his message was clear, as his candour in demolishing some of his own submissions was impressive.

There is a risk of oversimplifying his conclusions, but here are some of the main points in his closely reasoned thesis.

Three recent developments combine to make a re-examination of policy for the distribution of the national income overdue.

First there is (in Britain) the incomes policy itself, which has so far been little more than an emergency response to balance of payments difficulties caused by money wage and price inflation. As such it makes extremely heavy demands for voluntary restraint in the exercise of economic power by enterprises and labour unions, and it is highly unlikely that these groups will give up more than temporarily the objectives for which they believe they exist, except under one or the other of two conditions. One is compulsion, and the other is the formulation of much clearer objectives for the longer run. What is felt to be a purely negative short-term restraint would be more acceptable if the two sides could see and agree upon a goal to which it was eventually leading. This requires a much broader conception of incomes policy than we have at present.

Secondly, there is the possible effect on the distribution of income of accelerated technical progress. In popular usage most kinds of technological innovation are regarded as being labour saving and this conception is a perfectly correct one provided we add that most of them have been capital saving as well.

In practice there has been some substitution of capital for labour superimposed on these effects, as the capital stock has grown much faster than the labour force. The net result has been that the ratio of the national capital stock to the national income has stayed roughly constant or tended to drift downwards only slightly in the United Kingdom and United States during the present century. It suggests that in the historical context of the fits and starts of new investment opportunities opened up by the access to new markets and technical discoveries, the supply of risk capital and enterprise has been quite elastic provided it has been able to get an economic return of the order of about 10 per cent.

It does not follow from the fall of labour's share in the national income that there will also be growing unemployment. The conclusion is rather that full employment may only be maintained by allowing the real wage to rise relatively slowly, with adverse consequences for labour's share. This consideration takes the discussion of incomes policy firmly into the field of property ownership, and it is right that it should do so. In the United Kingdom at the present time there is restraint both on pay increases and dividend increases, but the latter has encouraged a rise in share prices which is not compatible with the ordinary stock market conditions by themselves. It gives the shareholder a means of postponing rather than forgoing his income gain, which is not available to the employee.

Thirdly, there is the belief of some European labour leaders that there may be a "middle way" between *laissez-faire* capitalism and complete socialization with regard to the ownership of property no less than with the rise of prices as an integral part of decentralized planning.

Professor Crossley recalled: "It was a central proposition of the socialist revolutionaries of the nineteenth century that the possession of private property carried with it economic power and much of Marx's commentary on conditions as he found them shows that it was, indeed, used to exploit. But if we compare the actual structure of economic power as it now exists in those countries which abolished private property in a socialist revolution with the situation in a country like the United Kingdom, which did not, we find a surprising similarity. In Britain the interlocking structure of company boards, with some directors sitting simultaneously on the boards of several companies, their merchant banks and commercial banks, is not very different from the structure of planning committees in the USSR. In each case the primary objective is co-ordination and much use is made of financial control. There is continual reshuffling of resource allocation (increasingly, through take-overs and mergers in the United Kingdom) but the few hundred key co-ordinators remain the same whether they are company directors in the one case or state officials in the other. The similarity is strengthened by the recent growth in the United Kingdom of joint planning boards which encourage the pooling of information, expectations, and plans, and where trade union members sit with representatives of management and government.

"It would be wrong to push the comparison too far, because the directors of British companies do have shareholdings themselves. Some indication of the disparity in the size of the holdings is available in the reports of the Inland Revenue Commissioners, which suggest that more than half of the personal holdings of British ordinary shares quoted on the Stock Exchange are held by less than 5 per cent of the population. An increasing proportion of shares is held by institutions, but these are also being brought within the system of interlocking directorships; insurance companies, for example, often share directors with enterprises in which they have invested their funds, and there are also links with the unit trusts through the commercial banks which act as their trustees. There has also been a growth of direct small shareholdings, but it is likely that the diffusion of ownership in small units has strengthened rather than diminished the effective powers of the few who hold large stocks."

And then Professor Crossley examined some of the investment pay schemes which exist. Most are still at the discussion stage, and only in the Federal Republic of Germany has there been implementation on a significant scale. He added: "It is true that for several years a number of US companies have operated wage-savings schemes which are usually called Thrift Plans, and some of the large British companies, like ICI Ltd., have specially favourable arrangements whereby employees can obtain shares in their own company. But the proposals, under consideration in Germany, the Netherlands, and Italy, and to a more limited extent in the UK, have a considerably wider scope. It is a common element in them that some part of the employee's pay is set aside contractually, and used for accumulation of assets on his behalf."

He continued: "A first important distinction is between schemes set up by the government, by employers acting unilaterally, and by joint action through voluntary collective bargaining. One example of the first type is the proposal for a save-as-you-earn scheme recently made by the British Conservative Party, which appears to be very similar to the arrangements which have existed in Germany under the House Savings Law of 1953 and the Premium Savings Law of 1959; the first offering the choice between a premium or a tax and social security payment concession on savings deposited by individuals in respect of intended house purchase and blocked for 5 years, and the latter giving a bonus of 20 per cent on the interest earned on approved types of savings for the same period. This figure has recently been raised to 25 per cent. The tax and social security payment concession and the interest premium were brought together in further legislation in 1961 to permit wage-saving schemes of the second type, i.e. arrangements offered unilaterally by employers (but under state sponsorship and subsidy in this case)."

Professor Crossley explained that the individual worker's participation was voluntary, but the benefits were restricted to savings not exceeding 312 Deutschemarks a year. He added: "Since the net return was very high in relation to the small size of the outlay (tax and social security contributions amounting to about one-third of the wage) it might have been expected that many workers would take advantage of the scheme. But according to a survey made by the Federal Statistical Office in 1963, only 2 per cent were participating, and the greater part of their apparently 'extra' savings was in fact a substitute for other fringe benefits which they gave up. In the Netherlands the participation is much higher in the similar savings schemes which are organized at works level, even though the direct tax advantage to the employee is less. This may be because the Dutch government provides a variety of premia and tax concessions on savings to various classes of individuals in any case, independently of the works savings schemes. In the United Kingdom, the Confederation of British Industry has proposed that schemes of this second type should be set up, with employees specifying contributions voluntarily, up to a maximum of, say, 10 per cent of earnings. There is no reason for doubting their argument that some means such as tax concessions will be needed to raise quite substantially the return on small savings, if the scheme is to succeed, but the government have not yet stated their views."

It was the disappointing result of the DM 312 Law in Germany, coupled with a particularly strong feeling there that property was very unevenly distributed, which prompted the most widely discussed proposal for a collectively bargained investment pay scheme to have appeared so far; the so-called Leber Plan. Georg Leber, the president of the Construction Workers Union, has been an outstanding innovator in industrial relations in post-war Germany.

Leber stated his objectives: "The existing wages policy can do nothing to change today's property structure in our nation. If we are to improve this structure we must abandon the policy of paying mere wages and proceed to a comprehensive income policy. It is not enough simply to increase that part of the workers' income directed to satisfying their consumption needs; we must also place wage earners and salaried employees, i.e. the workers, in a position where they can share in the capital assets of industry, and thus in the ownership of capital goods. These are the three aims of our scheme: (1) the workers shall receive a fairer share of the national product; (2) such sharing will enable them to participate in economic growth, and will help to strengthen their sense of acceptance as full citizens; (3) since under the scheme the investment ratio needed to promote industrial growth remains secure, the proposed method of accumulating assets will concurrently contribute to the stabilization of the country's currency."

Although the details of the original Leber Plan were not all implemented in the Construction Workers' Agreement which was signed in March 1965, their wide public discussion in an election year led to major improvements in the DM 312 Law. Although employers in other industries were hostile, and the Federation of German Trade Unions took up a cautious attitude, more agreements along the lines of the Leber Plan now seem likely in Germany.

Professor Crossley produced an exhaustive economic evaluation of these investment-savings schemes, pointing out that such new channels might merely switch savings from existing channels rather than increase the total flow. He discussed the advantages and disadvantages of a system of compulsory savings, the immense importance of increased productivity to maintain interest rates and increase national wealth, and the danger that these schemes might inhibit the mobility of workers in so far as they may not be transferable between employers. He conceded, however: "Efficiency on the job may be improved both through the direct interest which the employees have in higher profits, and more generally through a greater identity of purpose between management and employees. There is little difference here, though, from ordinary profit-sharing schemes and these give only limited grounds for optimism."

He admitted: "The prospects for a rapid redistribution of the ownership of wealth by means of investment pay schemes do not appear to be high. This is reinforced by a consideration of the magnitudes involved. In the United Kingdom the total value of industrial ordinary shares quoted on the Stock Exchange already exceeds the annual wage and salary bill, and it would be many years before an investment pay scheme (particularly a scheme of the incremental type) could increase the holdings of employees sufficiently for share price appreciation to provide much compensation to them for the unequal effects of pay and dividend restraint. But other economists may disagree with this pessimistic short-term outlook, and it is well to recall that human motives are fortunately not governed wholly by the conclusions of economic analysis. This is especially so with the issue of property ownership and the distribution of incomes, which is best seen in a broader, long-term social and political context. Firstly, it is a matter of great importance for employees and employers whether investment pay schemes should be introduced by collective agreement, if they are introduced at all. This resolves mainly into the question whether trade unions believe that participation in the design and administration of such schemes is a proper activity for them. Secondly, if we are right in thinking these schemes will not redistribute property quickly enough to meet growing pressures for greater equality, there is the question of appropriate government action, possibly of a quite radical kind towards the same objective."

Professor Crossley declared: "The arguments for and against collectively negotiated schemes are partly practical and partly matters of principle. At the practical level, trade unions cannot count on the support of sympathetic government policies continually, however helpful they may be from time to time. In addition, it is unlikely that schemes offered unilaterally by the employers will always take the form most wanted by the workpeople, and this may differ between companies and industries according to such factors as the stability of pay and employment levels. If there is to be a departure from individual freedom of choice to participate, the only conceivable alternative would be a contract signed by the union on the basis of a democratically determined union policy. It is in addition much more likely that the beneficial effects on productivity stressed by some advocates of investment pay schemes will be achieved where the two sides come together in collective bargaining. The example of productivity agreements in the United Kingdom is relevant here, though these have not so far involved much departure from conventional methods of payment. The main practical arguments against collectively negotiated schemes would be an attempt by trade unions to add the investment wage on to the wage increase they would normally have sought. Employers are only too likely to regard this as a bid for a disguised extra wage increase, and if the concession is made, the consequences will be cost inflation, possibly an increase in unemployment and a withdrawal of capital. None of this need happen if investment pay schemes are treated as a contractual allocation of normal wage increases, and at the practical level, the main question to be faced by trade unions is whether they want that."

He continued: "On matters of principle there are undoubtedly many trade union leaders in the United Kingdom, France, and Sweden (and, indeed, in Germany, Italy, and the Netherlands) who oppose such schemes on the traditional ground that participation in management, in any form, is inimical to the free and independent pursuit of trade union objectives. For a minority, the stronger view is expressed that such schemes would imply identification with the capitalist form of society which they do not accept, and a further step (along with participation in indicative planning and various joint procedures) and a highly dangerous one, towards corporatism. These groups reject workers' participation in favour of workers' control, and it may be noted parenthetically that there is no indication that the neosyndicalists have seen the accumulation of strategic shareholdings as a possible means to achieve this end. But for a majority who do not seek an immediate revolutionary change, the belief that investment pay schemes would encroach on trade union freedom is, in the author's opinion, a mistaken one.

"Capitalist societies", added Professor Crossley, "are amenable to continuous change, and if they are to be changed to the advantage of the employee, property ownership, especially in the form of industrial shares, is a most important lever of power which can be used to this end. It would also be in the interests of society as a whole to see some decentralization of effective voting rights through the creation of several large trade union funds of the investment trust variety, bringing together the voting strength of many small owners and activating it in the pursuit of trade union objectives. To those outside the labour movement who fear the powers of this new kind of institution, it might be pointed out that participation in the ownership of capital also brings new responsibilities and the need for better understanding of the way our economies work, especially at the macro-economic level. This is already evident in the extremely high level of sophistication of the arguments used by those European labour leaders who have argued for what they rightly call a 'full' incomes policy."

Professor Crossley also discussed the forms which governments might follow if they were eager to stimulate greater equality in the ownership of wealth.

He argued that there was no incompatibility between collectively bargained investment pay schemes and substantial subsidies to small savers who enter into contractual arrangements to save, irrespective of the form of the contract. A more radical development would be the introduction of a capital tax, which would have the advantage, unlike a progressive tax on investment income, that it would discriminate less against personal investment in a high-yield form.

Professor Crossley continued: "The arrangements for taxation of properties at death vary considerably between countries, but in general they are not successful as a redistributive device if property which is given *inter vivos* is able to escape taxation. There is much to be said for abolishing the system of death duties in the United Kingdom, for example, and replacing it by a progressive capital receipts tax on the accumulating property received by persons from all gifts throughout their lives. Secondly, and again as part of a policy of seeking a 'middle way' between complete *laissez-faire* capitalism and full socialization, the government could both reduce the proportion of property in private hands and exercise a strategic influence on industrial structure and policy by greatly increasing the number of industrial shares which it holds and uses on the taxpayers' behalf."

He mentioned that in the United Kingdom the government at present holds about 2 per cent of quoted shares. A major obstacle to this policy, he conceded, is the existence of a large national debt not wholly backed by real assets. Taking the balance of the financial liabilities against real assets sector by sector, Mr. Jack Revell has estimated the distribution between sectors of the national net worth, and calculates that the public sector holding was a negative figure of – 10 per cent in 1958, a condition which Professor Meade has described as the "anti-socialist state".

Professor Crossley suggested "A policy of retiring the national debt and using the proceeds to acquire equity holdings would require the combination of a tight fiscal policy—a budget surplus—with a relatively easy monetary one, since bond prices would tend to rise and depress the long-term rate of interest. An easy monetary policy is contrary to current orthodoxy, but there are distinct advantages for real investment outlay in sustained lower interest rates, and in the United Kingdom at least, the support for the bond market would be very welcome."

He concluded: "Finally, few governments have yet begun to comprehend the full scope of their powers and responsibilities with regard to what is probably the most important single factor influencing the sharing of income, which is the distribution among the population of human capital in the form of education and training."

He explained: "Since the level of savings rises with income, there is a strong force linking high earning power with the accumulation of property, so the distribution of education underlies the distribution of wealth as well. This is reinforced by the social stratification which accompanies inequalities in educational opportunity and attainment, which leads on the one hand to the further concentration of properties through marriages, largely within each social class, and also to the reinforcement through the lower fertility rates among those with large properties, since high birth rates tend to be correlated with low educational levels, no less between families than between countries."

In opening the discussion at the conference, Professor Crossley observed: "One of the few precepts which has survived our experiments with economical policy in the United Kingdom in the last decade is the old fallacy that 'if it hurts, it must be doing some good'. While I have to report that by that criterion we now have a very successful incomes policy in the United Kingdom—it certainly hurts. And it is little comfort to those on the receiving end that the government can point out, in the words of another old saying, that it hurts them at least as much as it hurts us."

But he does not think that Britain's incomes policy will go away. He added: "It is here to stay in one form or another. I don't think we can escape the challenge it presents to social ingenuity, a challenge ultimately to our political maturity."

It was Professor Crossley who calculated for the Prices and Incomes Board in 1968 that the effect of the incomes policy had been "to reduce the rate of increase of employment incomes by something of the order of 1 per cent a year". Some critics had said that if the effect was as small as that, the policy was not worth all the effort.

But Professor Crossley explained: "It is true that one percentage may not seem a great deal, but, of course, it is one percentage point out of five or six percentage points a year—that is to say a reduction in the rate of inflation of the order of 15 or 20 per cent, and that is not a trifling achievement."

The alternative to an incomes policy would have been an unemployment rate of 3 or 4 per cent.

Professor Crossley explained that his paper had attempted to do three things. One was to bring together the variations in the existing schemes of investment pay. The second was to show how important it was to consider this idea seriously. And third, to see if these schemes had the effects which had been claimed for them.

His own preference is for collectively negotiated schemes which have participation in equity shares, which both accumulate in value and carry voting rights.

But he conceded, with an apology for being pessimistic, that unions were unlikely, by acquiring shares, to get a greater total compensation than by the traditional method of negotiating a straight wage increase, and it was improbable that the acquisition of shares would add substantially to the proportion of the national income which is already saved.

Nevertheless, despite the candid qualifications of his advocacy, Professor Crossley stimulated plenty of discussion and provoked more reservations from the conference participants.

Particularly telling comments came, for example, from Mr. Douglas Taylor, the deputy director-general of the Confederation of British Industry. He recalled that Professor Crossley had argued that there were many constant relationships in his economic model; a constant return on capital of about 10 per cent; a constant relation between capital and output; a constant share of the capital income going to wage-earners; a constant level of savings within the economy.

The main theme, therefore, was that there is nothing that management or labour can do to affect these relations in the longer term. It follows that schemes to enable workers to acquire capital cannot make much difference to the workers' share of the income. Mr. Taylor continued: "If we accept this general framework then the whole operation of the market, and in particular, of the collective bargaining process is quite ineffective in influencing these matters."

He declared: "Our experience as British management is that in the all important short term it is not true that collective bargaining is ineffective in these matters. It is not true that the return on capital is constant, or that the workers' share of total incomes is not rising. It has become more and more difficult to hold wage rises in check or to limit consumption, and the return on capital has steadily fallen. It seems to me that Professor Crossley's paper pays too little attention to the amount of capital held on behalf of workers by institutions. First of all, we have to think of the enormous holdings of the Pension Funds. These are well over £8,000 million. One of the main themes of my Confederation's activity at the moment is undertaken by a team led by Mr. Kerr-Muir, who is here, to persuade the chancellor and the government not to do something in the new pension plans which is going to affect these very large holdings."

Mr. Taylor continued: "A second main holding by institutions, of course, is in the insurance companies and in the building societies for house purchase. And, thirdly, we have the direct holdings of the trade unions. Here we have both union holdings and the Trade Union Unit Trust holdings, a development particularly associated with Lord Hirshfield, the secretary-treasurer of the British delegation at this conference."

Mr. Taylor disclosed that his Confederation is trying to devise a scheme for workers acquiring capital in industry. He commented, from his own investigations, on four of the issues raised by Professor Crossley.

(1) This was the idea of the collective bargain which gives certain increases in cash, and in addition, certain increases by way of holding shares which have to be locked in for a period.

Mr. Taylor commented: "We came to the conclusion that this was not a runner in the current situation in the United Kingdom, for two reasons. The first is that any provision for the worker to hold shares as the result of collective bargaining would be additional to the cash benefits of the collective bargain, and not in substitution for part of them. We thought long over this, but the whole of our advice from our members was that this would be the case; and the whole of our experience is that what the unions look at is the take-home pay in their wage packet. They are not interested in determining the size of the bargain they will accept in these postponed benefits. The single reason for rejecting the negotiated schemes is that there is the most enormous administrative complexity. You have got to remember that it affects not only the people who are engaged in a particular plant or industry at the time, but those who are going to come into it, and those who are going to go out of it with locked-in provisions. So you have enormous demands for records and following up these various people. And in the United Kingdom we have the greatest labour mobility in the sense of job mobility. You know that we have more than 8 million job changes a year in a labour force of some 25 million."

He continued: "Our simple conclusion was that you could not support, on incomes policy grounds, negotiated pay schemes because they would be additional to settlements on the incomes policy, and the economy could not afford those additions."

(2) Support could be given to certain types of schemes with a savings bias on other economic grounds, which were related.

Mr Taylor explained: "We have a great need in our country for additional savings. We have great threats to our existing savings, which are represented by inflation because the returns on some types of savings are almost negatived. But the fact is that the moment you begin to improve the returns on any particular saving, you do not increase the total volume of saving, if it is possible to switch from the less attractive to the more attractive stock. But a voluntary collective savings scheme of a negotiated character, a contractual scheme, has the particular benefit of minimizing the possibility of switching because it is largely attracting into the savings field the workers and people who have not easily switchable savings at their disposal."

(3) Any scheme must be voluntary. This means that there must be some substantial incentive to lead the worker into such a scheme. The most suitable incentive is probably a very considerable tax advantage.

(4) This conclusion is in line with Professor Crossley's—that such schemes can have only a marginal effect. Their magnitude cannot be very great, or their administration costs would be formidable.

Mr. Taylor revealed that an "exercise" had been undertaken on administrative costs, and this suggested that apart from the employers' own costs of deduction, the central cost of a "comparatively crude scheme" would approach 20 per cent.

He disclosed: "We have put a scheme on these lines to the government. We have been negotiating about it on a confidential basis for about 9 months, with the benevolent neutrality of the TUC. It is a scheme for deductions of up to £3 a week from workers' pay which would be locked in for a period of 5 years, with limited provisions for withdrawal on hardship grounds, and with a substantial tax relief of at least 50 per cent of the sum saved. That is, tax relief not on the final benefits, but on the actual sum saved. There would be an option for the employee to make one of three choices into which the savings would go; either a new government security of a much more attractive character, possibly index-linked, or a unit trust, which would invest in equities, with which we hope to see both CBI and TUC sponsorship. And, thirdly, there would be some form of building society investment."

Mr. Taylor continued: "In the negotiations with the government it was clear that they saw two very serious difficulties. One was administrative costs, and the second was the effect on the equity market driven up by general inflationary tendencies. The government were reluctant at this stage to contemplate such a fund going into equities."

Although he agreed with many of Professor Crossley's conclusions, Mr. Taylor "strongly rejected" the inference that the government might do something in the way of income redistribution through, perhaps, a wealth tax.

"I would suggest", Mr. Taylor urged, "that we should regard a wealth tax as completely inconsistent with the aims of an incomes policy for general economic needs. A wealth tax would, in fact, reduce savings and would increase consumption. Our inference is that a contractual savings scheme is in line with an incomes policy."

Lord Peddie, a deputy-chairman of the Prices and Incomes Board, doubted Professor Crossley's assumption that there would be an automatic and steady progression in the value of capital in relation to labour. It was likely that labour, and particularly the higher quality of labour, would secure an increasing proportion of the national income.

Nor did Lord Peddie accept that a prices and incomes policy was a permanent restriction on negotiations. He added: "I think the operation of a prices and incomes policy, and the work of the Board, has introduced a new dynamism into negotiations. I believe, for instance, that the development of productivity agreements will create far more sophisticated negotiations with encouragement of technological developments and the securing of even better returns."

Mr. Yosef Ami (Israel) was attracted by Professor Crossleys' theme because the investment of workers in their own plant would stimulate an interest in its efficiency and profitability, and would deepen the common interest of management and labour to solve other problems.

Mr. John Peel, one of the leaders of the Yorkshire textile workers and a member of the TUC General Council, accepted that a prices and incomes policy had restrictive connotations. He was interested in the alternatives in different forms of savings, and thought that, quite apart from the mechanics of these schemes, reasonable human ingenuity could overcome their drawbacks. But he thought Professor Crossley had "glossed over" one or two difficulties. He seemed to take it for granted that workers wanted to save.

Mr Peel observed: "There is a difficulty in persuading people to save even when the facilities offered are attractive. As for the incomes policy, its feature isn't just a question of restraint as much as its redistributive element. It has been said recently that the top eight of our labour forces in England earn twice as much as the bottom eight. Incomes policy in England isn't breaking down just because of its restrictive connotations, important though they may be, but because of the inherent injustice which workers see in the policy, operating as it does in a wages structure which is completely archaic."

Inevitably, Professor Crossley's suggestion of similarity in the "seats of authority" between Britain and Soviet Russia was bound to provoke some reaction, though it was not raised until the concluding session of the conference.

Mr. Taylor recalled that the Professor had drawn attention to the system in Britain of the interlocking structure of company boards, with directors sitting simultaneously on the boards of several companies, their merchant banks and commercial banks, and compared it with the planning system in the Soviet Union. He said that there is continual reshuffling of resource allocation, increasingly through takeovers and mergers in the United Kingdom; that the few hundred key co-ordinators remain the same, whether company directors or planning officials.

"I think this is not in line with my own experience", Mr. Taylor continued. "I would be rather surprised to think that it was so widespread in our country, particularly in the largest companies, the ICIs, the Unilevers, and so on. I would have thought that it was of importance, perhaps for efficiency rather than planning, and that from the standpoint of efficiency it is perhaps commoner in the United States than in the United Kingdom."

And Mr. R. J. Kerr-Muir, a director of Courtaulds Ltd., declared: "In the United Kingdom this matter of interlocking executive directors just does not happen. In the United States it does to a certain extent, but not very much on the executive side. As far as my company is concerned, we have a number of former members of governments who come in; we have two ex-presidents of the Board of Trade, we have an ex-minister of education. And we do find that, on our Board, the contributions which they make to the discussions are enormously useful from the point of view of the attitudes which we take towards human relationships within the company."

But Professor Ben Roberts thought that Professor Crossley's concept "is an extremely important one, and there is some relevance in it. I would put it rather differently from the way he put it."

Professor Roberts continued: "What impresses me is that, in the very many conferences and management educational activities that I take part in, it is the same people, more or less, who come to these activities over and over again. You can say that the dynamic in British industrial life is coming from about 200 companies;

it is their directors, their chief managers, whether at the personnel level, or the production level, or the level of finance director, whatever it is, that are really making the running in this respect. And I think this is a very important factor in the whole machinery that we have established in Britain of committees, and so on, to work with government in promoting economic development, technical change and so on."

He added: "In planning terms, it is this group that is significant, and I think to this extent there is a certain parallelism with the Soviet Union, although the system is quite different and it is worked out in a different way. Now, I think this is true of other countries, and the point has been brought out very specifically in the case of France. If you read the studies that have been made of the French planning system, this is being articulated and formulated quite precisely; this is the way in which the French planning system works, through a high level of the top companies, a group of people who all went, broadly speaking, to the same schools—it is the French equivalent of our public school system, in effect—the same group of people are in the planning ministries, and so on. You get that kind of interrelationship."

Professor Roberts continued: "What does this do? It provides a common understanding; it provides a common language; it provides, as it were, a common status level, so that communications and interrelationships are possible. It makes for communication, which is extremely important, between government and industry. In other words, it is sometimes called the Establishment; this is the Industrial Establishment, and I think that it is extremely important that the trade unions are a part of this. They are in exactly the same situation."

He explained: "If you look at the phenomena of the trade union movement in terms of its behaviour, it is exactly the same. If you go to trade union conferences, conferences that are organized by the TUC on educational activity, it is a certain group of unions who are represented every time. A large number of unions never send anyone.It is an institutional, behavioural phenomenon, and it is extremely important. It is related to size, power, and so on."

Professor Roberts concluded: "And so I would say that there is a good deal in what Crossley was saying here that makes it highly significant from the point of view of planning. The problem, however, is that we are failing to communicate; we are failing to put across our message to that large part of the economy—the smaller company which is generally less efficient, the smaller union which is generally less efficient, and somehow we have not been able to achieve any means of communicating, or any means of influencing, that factor. This is a major problem, it seems to me, from the point of view of democratic planning in our modern communities."

But the final tribute to Professor Crossley came from Mr. Theodore Kheel in the official summing-up of conference proceedings.

Mr. Kheel concluded: "He made an excellent analysis of our ability to increase our productive capacity, and also of our inability to devise a system of distributing the fruits, the income of our productive capacity, on an equitable basis. He traced the relationship of the investment of capital and labour and observed that they had remained relatively constant over the last 50 years. He made the important observation that, with automation, the relationship between capital and labour is likely to change. I think the consequences of this are so far-reaching that further study is clearly indicated."

Appendix A

International Conference on Technological Change and Human Development, Jerusalem, 14 - 18 April, 1969

Co-sponsored by: Foundation on Automation and Employment Ltd. (Britain).
American Foundation on Automation and Employment, Inc.
Government of Israel; Federation of Labour (Histadrut); Manufacturers' Association of Israel; The Hebrew University of Jerusalem; Haifa Technion; and the Israel Institute of Productivity.

Monday 14 April

Technology and the Human Need for Economic Growth

Session
CHAIRMAN: THE RT. HON. LORD CARRON, KSG, MA, DSc, Director, the Bank of England, and Former President, Amalgamated Engineering Union.

No. 1: *The Role of Modern Technology in Raising the Economic Standards of Less Developed Countries.*

Author: PROFESSOR NICHOLAS KALDOR, MA, Professor of Economics, University of Cambridge.

Commentator: C. HENNIKER-HEATON, CBE, Director, British Spinners' and Doublers' Association.

No. 2: *The Role of Technology in the Development and Viability of Israel.*

Author and Commentator: PROFESSOR S. IRMAY, Technion—Israel Institute of Technology.

No. 3: *Technology and the Dispossessed.*

*Author:** WHITNEY M. YOUNG, JR., Executive Director, National Urban League, Inc.

Commentator: ISADOR LUBIN, PhD, Economic Consultant, The Twentieth Century Fund.

Tuesday 15 April

Employment Problems in an Automated Society

Morning Session

CHAIRMAN: PROFESSOR JAY TABB, Technion—Israel Institute of Technology.

No. 4: *Changing Skill Requirements, Job Training, Motivation and Mobility.*

Author: BEN-AMI FRIEDRICH, DIP ENG, Member, Executive Committee Manufacturers' Association of Israel.

Commentators: YOSEF AMI, Vice President, Technion—Israel Institute of Technology.

DAN RAM, Assistant for Vocational Education, the Ministry of Labour.

No. 5: *The Reorientation of Attitudes of Employers, Trade Unions and Employees to Changing Job Content.*

*Author:** SIR DENIS BARNES, KCB, Permanent Under-Secretary of State, Department of Employment and Productivity.

Commentator: DAVID BASNETT, National Industrial Officer, General and Municipal Workers' Union.

Afternoon Session

CHAIRMAN: ARIE GURAL, CE, Director, Ministry of Labour.

No. 6: *Youth, Motivation, Jobs, and Technology.*

*Author:** KENNETH B. CLARK, PhD, President, Metropolitan Applied Research Center Inc.

Commentator: DAVID G. MOORE, PhD, Dean, New York State School of Industrial and Labor Relations, Cornell University.

* In the unavoidable absence of: Mr. Whitney M. Young, his paper was presented by Bayard Rustin, Executive Director, A. Phillip Randolf Institute, Inc.; Sir Denis Barnes, his paper was presented by Frank Pickford, Assistant Under Secretary of State, Department of Employment and Productivity; and Mr. Kenneth B. Clark, his paper was presented by James R. Dumpson, PhD, Dean, Fordham University School of Social Service.

Wednesday 16 April

Technological Change and Industrial Relations

Session

CHAIRMAN: BERTRAM A. POWERS, President, New York Typographical Union No. 6.

No. 7: *The Future of Collective Bargaining and The Role of Industrial Conflict.*

Author: P. L. SIEMILLER, International President, International Association of Machinists and Aerospace Workers.

Commentator: E. T. KLASSEN, the Deputy Postmaster-General and Former President, American Can Company.

No. 8: *Incomes Policy and Sharing in Capital Gains.*

Author: J. RODNEY CROSSLEY, BSc (ECON), Professor of Industrial Relations, University of Leeds.

Commentator: DOUGLAS TAYLOR, CMG, Deputy Director-General, Confederation of British Industry.

No. 9: *Workers' Participation at the Plant, Union and National Level in Israel.*

Author: COLONEL ISRAEL MEIDAN, MBA, Director-General, Israel Institute of Productivity.

Commentators: JOSEPH HAUSMANN, LLM, Director, Department of Labour, Manufacturers' Association of Israel.
CHAIM KOHN, Director of Productivity Division, Trade Union Department, Histadrut.

Thursday 17 April
Corporations, Trade Unions and the Individual in the New Environment

Session CHAIRMAN: HAROLD A. WEBSTER, Chairman of the Board Emeritus, T. Frederick Jackson Inc.

No. 10: *The Role of the International Company.*

Author: PETER PARKER, MVO, Chairman, Bookers Engineering & Industrial Holdings Ltd.

Commentator: EDWIN R. NIXON, Managing Director, IBM, United Kingdom Ltd.

No. 11: *The Social Role of the Unions.*

Author: YERUHAM MESHEL, Chairman, Trade Union Executive Department, Histadrut.

Commentator: RIVKA BAR-YOSEF, PhD, Head, Department of Sociology, the Hebrew University.

No. 12: *Artists, Engineers and Technology.*

Author: J. WILHELM KLUVER, PhD, President, Experiments in Art and Technology Inc.

Commentator: EDWIN F. SHELLEY, President, E. F. Shelley & Co. Inc.

Panel Discussion and Conference Summing Up

Session CHAIRMAN: T. LESLIE BLAKEMAN, CBE, Director, Labour Relations, Ford Motor Co. Ltd.

Panel:

B. J. A. HARGREAVES, Director of Public Affairs, IBM, United Kingdom Ltd.

MATTHEW A. KELLY, PhD, Professor of Industrial Relations, New York State School of Industrial and Labor Relations, Cornell University.

MARY CONWAY KOHLER, Director, National Commission for Resources for Youth Inc.

COLONEL ISRAEL MEIDAN, MBA, Director-General, Israel Institute of Productivity.

JACK PEEL, General Secretary, National Union of Dyers, Bleachers, and Textile Workers.

PROFESSOR BEN R. ROBERTS, MA, Professor of Industrial Relations, London School of Economics and Political Science.

PROFESSOR JAY TABB, Technion—Israel Institute of Technology.

HARRY VAN ARSDALE, JR., President, New York City Central Labor Council AFL-CIO.

Summing up:

THEODORE W. KHEEL, Battle, Fowler, Stokes, and Kheel.

Appendix B

List of Participants at the Conference

British

Trade Union

Basnett, David. National Industrial Officer, General and Municipal Workers' Union.

Davies, Dai. General Secretary, Iron and Steel Trades Confederation.

Fisher, Patrick. Head of Production Department, Trade Union Congress.

Peel, Jack. General Secretary, National Union of Dyers, Bleachers and Textile Workers.

Plant, Cyril T. H., OBE. General Secretary, Inland Revenue Staff Federation.

Smith, G. F. General Secretary, Amalgamated Society of Woodworkers.

Tennant, Douglas, CBE. General Secretary, Merchant Navy and Airline Officers' Association.

Government

Pickford, Frank. Asst. Under-Secretary of State, Department of Employment and Productivity.

Skinner, O. J. Labour Attache, United Kingdom Embassy, Tel Aviv.

Management

Blakeman, Leslie T., CBE. Director, Labour Relations Ford Motor Co. Ltd.

Briggs, Harry, CBE. Confederation of British Industry.

Cartwright, W. F. Managing Director, British Steel Corporation, South Wales Group.

Farrimond, H. L. Personnel Director, Upper Clyde Shipbuilders Ltd.

Hargreaves, B. John A. Director of Public Affairs, IBM, United Kingdom Ltd.

Henniker-Heaton, C., CBE. Director, British Spinners' and Doublers' Association.

Kerr-Muir, R. J., OBE. Director, Courtaulds Ltd.

Nixon, Edwin R. Managing Director, IBM, United Kingdom Ltd.

Parker, Peter, MVO. Chairman, Bookers Engineering & Industrial Holdings Ltd.

TAYLOR, DOUGLAS, CMG. Deputy Director-General, Confederation of British Industry.

WHEELER, SIR CHARLES, KBE. Director, Guest Keen & Nettlefold Ltd.; Director, Rudolf Wolff & Co.

Academics

CROSSLEY, PROFESSOR J. R., BSc (ECON), Professor Industrial Relations, University of Leeds.

KALDOR, PROFESSOR NICHOLAS, MA. Professor of Economics, Kings College, Cambridge; Special Adviser to the Chancellor of the Exchequer.

ROBERTS, PROFESSOR BEN R., MA. Professor of Industrial Relations, London School of Economics and Political Science.

Others

CARRON, THE RT. HON. LORD, KSG, MA, DSc. Director, the Bank of England; Ex-President, Amalgamated Engineering Union.

DOUGLASS, THE RT. HON. LORD, OF CLEVELAND. Member, Electricity Council; Member, Monopolies Commission, *ex* General Secretary.

EVANS, SIR TREVOR, CBE. Director, The Beaverbrook Press.

HIRSHFIELD, THE RT. HON. LORD, FCA. Chairman, Trades Union Unit Trust Managers Ltd.; Deputy Chairman, Northampton New Town Development Corp.

PEDDIE, THE RT. HON. LORD. Deputy-Chairman, National Board for Prices and Incomes.

WOODWARD, JOAN (Mrs. Blakeman). Reader in Industrial Sociology, Imperial College.

ARKELL, BETTY. Secretary, Foundation on Automation and Employment Ltd.

American and Others

Trade Union

DEMORE, MATTHEW. General Secretary-Treasurer, International Association of Machinists and Aerospace Workers.

FAUPL, RUDOLF. United States Worker Delegates, International Labor Organization.

JIRIKOWIC, VERNON E. Director of Research, International Association of Machinists and Aerospace Workers.

OTTLEY, PETER. President, Local 144, Hotel and Allied Service Employees Union.

POWERS, BERTRAM A. President, New York Typographical Union No. 6.

SIEMILLER, P. L. International President, International Association of Machinists and Aerospace Workers.

TERPE, KEITH. Executive Vice-President, Seafarers International Union of North America.

TOWNSEND, JACK. President, Bartenders Union, Loc. 15, New York City.

VAN ARSDALE, HARRY, JR. President, New York City Central Labor Council AFL-CIO

Management

HORWITZ, MARTIN. Chairman of the Board and General Counsel, United States Smelting Refining & Mining Co.

KLASSEN, ELMER T. The Deputy Postmaster-General; Former President, American Can Co.

LEEDS, LAURENCE C., JR. President, Manhattan Industries Inc.

WEBSTER, HAROLD A. Chairman of the Board Emeritus, T. Frederick Jackson Inc.

Academics

DUMPSON, JAMES R., PhD. Dean, Fordham University School of Social Service.

KELLY, MATTHEW A., PhD. Professor of Industrial Relations, New York State School of Industrial and Labor Relations, Cornell University.

MOORE, DAVID G., PhD. Dean, New York State School of Industrial and Labor Relations, Cornell University.

Others

COLE, GORDON H. Editor, The Machinist.

DAVIS, DANIEL S. Associate Director of Public Relations, National Urban League Inc.

DOUGHERTY, EDWARD G. Attorney, Brotherhood of Railway Carmen of America.

GORO, HERB, Photographer.

GRAY, BARBARA. Secretary-Treasurer, American Foundation on Automation and Employment Inc.

KADEN, LEWIS. Assistant to Mr. Kheel, Battle, Fowler, Stokes & Kheel.

KAHN, LOUIS. Secretary, E. A. Kahn & Co. Inc.

KAYE, SEYMOUR P. Manager of Industrial Relations, ITT Europe Inc.

KHEEL, THEODORE W. Battle, Fowler, Stokes & Kheel.

KLUVER, J. WILHELM, PhD. President, Experiments in Art and Technology Inc.

KOHLER, MARY CONWAY. Director, National Commission for Resources for Youth, Inc.

KRAMER, LEO. Chairman, Leo Kramer International Ltd.; President, Leo Kramer, Inc.

LEVIN, EDWARD. Director of Programs and Research, Institute of Collective Bargaining and Group Relations Inc.

LUBIN, ISADOR, PhD. Economic Consultant, The Twentieth Century Fund.

MCFADDEN, JAMES. Executive Director, Manpower Education Institute.

MCNALLY, CHARLES. Secretary, Allied Printing Trades Council of Greater New York.

MARTON, MICHAEL.

PLUNKETT, MARGARET L. Labor Attaché, United States Embassy, Tel-Aviv.

PUGH, DOUGLAS G. Program Adviser, Social Development the Ford Foundation.

RUSTIN, BAYARD. Executive Director, A. Phillip Randolf Institute Inc.

SCHMERTZ, ERIC J. Impartial (Public) Member, Office of Collective Bargaining, City of New York.
SHELLEY, EDWIN F. President, E. F. Shelley & Co. Inc.
TOWNSEND, EDWARD T. Labor Editor, Business Week.
VICTOR, J. Editor in Chief, American Labor Magazine.
YOUNG, WHITNEY M. JR. Executive Director, National Urban League, Inc.

Israeli

Trade Union

ALON, A. Secretary, International Department, Histadrut.
ARAD, A. Trade Union, Histadrut.
BAHAT, S. Director, International Department, Histadrut.
BARZILAI, Y. Economic and Social Research Department, Histadrut.
CHERMESH, R. Institute of Social and Economic Research, Histadrut.
JACOBI, G. Member of the Executive Bureau, Histadrut.
KISHALES, S. General Secretary of the National Union Workers, Aircraft Industry.
KOHN, C. Director of Productivity Division, Trade Union Department, Histadrut.
KOHN, S. National Union of Metal Workers, Histadrut.
MESHEL, Y. Chairman, Trade Union Executive Department, Histadrut.
SACHAR, B. Director of Education and Cultural Center, Histadrut.
SHEMER, A. Trade Union Department, Histadrut.
VILENSKA, E., MRS. Histadrut.
ZAN-BAR, S. General Secretary of the National Union of Metal Workers in Israel, Histadrut.
ZETELNY, S. Histadrut.
YUDIN, Y. Head of the Workers' Participation in Management Department, Histadrut.

Management

ALLON, Y. Director, Data Processing Division—Government and Government Institutions.
ALSHECH, J. Economic Department, Manufacturers' Association of Israel.
BAR-NETZ, M. Personnel Manager, Asbestos-Cement Industries.
BEN-ZEEV, Y. Director, TAAS.
ESHKOL, M. Director, Data Processing Division in Industry, IBM.
FRIEDRICH, B. A., DIP ENG. Asbestos-Cement Industries.
GADISH, F. Director-General Israel Chemical Fibres Ltd., Industrial Area.
GIDRON, A. Deputy Director-General, Electra (Israel) Ltd.
GOLDWASSER, A. Director-General, Oxidon–Israel Oxidation Co. Ltd.
HAUSMANN, J., LLM. Director, Department of Labour, Manufacturers' Association of Israel.
KAMINITZ, H. Administrative Director, Argaman Ltd. Textile Dye Works.

KARSZON, A. Managing Director, Vulcan Foundry Ltd.
KIZILSTEIN, M. Employment Services.
KLIER, S. Deputy Director-General, Argaman Ltd.—Textile Dye Works.
LIEBER, M. Director-General, The Lieber Co. Ltd.
MESSER, O. Director-General, Arad Chemical Industries Ltd.
MUSHIN, D. Director-General, Islatin Enterprises Ltd.
MODAY, Y. Director-General, Revlon.
POLLAK, D. Director-General, Polgat Woollen Industries.
REINER, E. Managing Director, Development Areas Industrialization Ltd.
SHELEF, Z. General Manager, Tamam.
SYRKIN, S. Consultant Engineer.
TAMIR, P. General Manager, Manufacturers' Association of Israel.
TOLKOWSKY, D. Director-General, Discount Bank Investment Corp. Ltd.
WEILER, J. V. Director-General, Elector-Optical Industry in Israel Ltd.
ZELKIND, A. Director-General, Elko Ltd.

Government

AVIGAD, M. Head, Vocational Training Department, Ministry of Education and Culture.
BAR-EL, D. Head, Management and Manpower Section, Israel Institute of Productivity.
BAR-ON, COL. I. Director, Department of International Co operation and Foreign Liaison, Ministry of Defence.
BERMAN, M. E. Director, Engineering Services, Ministry of Posts.
BEN-ISRAEL, G. Chief Labour Relations Officer, Labour Office.
CHOREV, A. Chief of Q Branch, IDF Headquarters.
DASH, Y. Head of Planning, Ministry of Interior.
DORON, G. Deputy Director, Israel Institute of Productivity.
FRENKEL, MAJOR E. Military POB 2770, Zahal.
GAL-ED, I. Director-General, Ministry of Development.
GERTZ, A. Adviser on Automation to the Ministry of Finance.
GEVA, R. Deputy Director-General, Ministry of Labour.
GILBOA, M. Assistant Director, Unit for Co-operation and Foreign Liaison, Ministry of Defence.
GINOR, DR. F. Economic Adviser to the Governor, Bank of Israel.
GOLDBERG, D. Head, Training Center, Israel Institute of Productivity.
GUZMAN, R. Budget Department, Ministry of Finance.
HOECHSTER, E. Director, Establishment and Planning Division.
JACOB, COL. I. Assistant Chief Scientist, Ministry of Defence.
KERNER, LT.-COL. Z. Military POB 2770, Zahal.
MAOR, N. Scientific Adviser, Ministry of Development.
MEIDAN, COL. I., MBA. Director-General, Israel Institute of Productivity.
MERHAV, DR. M. Ministry of Commerce and Industry.
MICHAEL, DR. A. Ministry of Labour.
NITZAN, J. Civil Service Commissioner.
RAAB, B. Head, EDP Section, Training Center, Israel Institute of Productivity.

RABINOWITZ, E. Head, Services Department, Israel Institute of Productivity.
RAM, D. Assistant for Vocational Training, Ministry of Labour.
RON-EL, A. Head, Systems and EDP Department, Israel Institute of Productivity.
ROSENWINE, B. District Inspector of Labour, Haifa and the North.
ROTTENSTREICH, Y. Head of Personnel, The Israel Institute of Productivity.
SAPHIR, J. Director, Technical Assistance Department, Prime Minister's Office.
SHA'ARI, Y. Deputy Minister, Ministry of Development.
SHABATAY, MRS. H. I/C of Apparel and Knitting Industry, Ministry of Commerce and Industry.
SHALON, U. Member of the Executive Committee, Israel Institute of Productivity.
SOLLY, S. Head, Organizational Behaviour Section in the Training Center, Israel Institute of Productivity.
STEIN, G. Scientific Adviser to the Ministry of Development.
URI, N. Labour Relations Officer, Ministry of Labour.
WEIL, L. Israel Standard Institute.
WISKIND, J. Israel Institute of Productivity.
WOLKOW, R. Deputy Director-General, National Insurance Institute.
YANU, S. Center for Policy Studies.

Academic

AMI, Y. Vice-President, Technion—Israel Institute of Technology.
BAR-YOSEF, DR. R. Head, Department of Sociology, The Hebrew University.
BEN-URI, PROF. J. Technion—Israel Institute of Technology.
GLOBERSON, DR. A. Tel Aviv University.
GRADITOR, D. Government Institute for Technical Training.
HARMETZ, J. Director-General, ORT Israel.
HOENICH, PROF. P. K. Technion—Israel Institute of Technology.
ILANY, DR. J. Head of Applied Research Department, Ministry of Commerce and Industry.
IRMAY, PROFESSOR S. Technion—Israel Institute of Technology.
ISRAELI, I. The Hebrew University.
LADANY, DR. S. Tel Aviv University.
MANNHEIM, DR. B. Technion—Israel Institute of Technology.
MESHULACH, DR. A. The Hebrew University.
OLLENDORF, PROFFESOR F. Technion—Israel Institute of Technology.
PERLBERG, DR. A. Technion—Israel Institute of Technology.
PERI, DR. J. N. Tel Aviv University.
SHAAL, G. Tel Aviv University.
SHIROM, DR. A. Tel Aviv University.
TABB, PROFESSOR J. Technion—Israel Institute of Technology.
ZANZIPAR, M. Technion—Israel Institute of Technology.

Index

Printed in Dunstable, United Kingdom

91342988R00117